Never A Dull Moment

My Life and Times

To Fred
with much affection

Geoffrey

Never A Dull Moment

My Life and Times

GEOFFREY WILSON

GeoffreyFawcettWilson.com

life
HIGHPOINT

Copyright © 2018 by Geoffrey Wilson

All rights reserved. Published in the United States of America. No part of this book may be reproduced or transmitted in any form or by any means, graphic, electronic or mechanical, including photocopying, recording, taping or by any information storage or retrieval system, without permission in writing from the publisher.

This edition is published by Highpoint Life.
For information, write to info@highpointpubs.com.

First Edition

ISBN: 978-1-64467-752-0

Library of Congress Cataloging-in-Publication Data

Wilson, Geoffrey
Never a Dull Moment: My Life and Times
Includes index.

Summary: "In this unique and fascinating autobiography, Geoffrey Fawcett Wilson documents his quintessential twentieth century life of service, adventure, scholarship and family in England and the United States." – Provided by publisher.

ISBN: 978-1-64467-752-0 (hardcover)
1.Biography

Library of Congress Control Number: 2018960327

Design by Sarah Clarehart

Manufactured in the United States of America

10   9   8   7   6   5   4   3   2   1

Contents

Preface		vii
Acknowledgments		ix
Chapter 1	1921–1931: The Early Years	1
Chapter 2	1931–1938: High School and Scouting	29
Chapter 3	1939–1945: The Second World War	57
Chapter 4	1945–1978: University Life	73
Chapter 5	1952–1953: Mountaineering and Marriage	105
Chapter 6	1955–1962: Pasture Cottage	133
Chapter 7	1960–1978: The Chamberlin Plan	151
Chapter 8	1948–1963: Conferences and Consultancies	181
Chapter 9	1962–1992: Woolwich Farm	199
Chapter 10	1992–2018: A New Life In a New World	223
Epilogue		239
Index		241

Preface

At first sight, this book may appear to be an autobiography, but it was never intended to be one. When forced to retire at the age of sixty after a catastrophic 'breakdown,' the medical consultants impressed upon me that I must 'forget the University altogether, and concentrate on the future'. So, I threw away all diaries and documents that might have reminded me of the past, and diverted all my thoughts accordingly.

More than twenty years later, and living in the United States, my daughter Susan judged that I was now recovered so well that it 'might not do any harm' just to write down some of the anecdotes I used to relate during her childhood, which she thought would interest my grandchildren.

So this volume began as a series of recollections cobbled together for the amusement of my family, but as I kept adding one episode after another, Sue, who is researching our family heritage, persuaded me to include such elements of my career and family history as it came readily to my deteriorating memory.

The comparison between life in the twentieth century and the present day proved to be irresistibly fascinating, and I'm deeply indebted to Sue for prompting me to embark upon this record of events which may not have any literary merit, but which has stimulated and intrigued me for some years now. I should like to 'dedicate' it—if that's not too pompous a word, with my love, to her.

G.W.

Acknowledgements

I am indebted to the following:

The Registrar of the University of Leeds for his courtesy in reviewing my first draft, and to the University for allowing me to reproduce the illustrations in Chapters Four and Seven.

My friend and colleague Fred Wilkinson, who examined and corrected my first draft with meticulous care and made suggestions to improve the text.

Avanti Architects and their photographer Tom De Gay for the picture of the front elevation of the Roger Stevens Lecture Theatre block on page 171.

Leodis for his picture of the Duchess of Kent, page 101.

Wikipedia Commons for making the illustrations on the following pages freely available for reproduction on pages 63, 64, 65, 67, 81, 94, 137, 171, 172, 191,194, and 198.

Paul Clements for his help and advice in formatting the text.

To Eileen Brill Wagner for introducing us to Michael Roney and his team at Highpoint Executive Publishing, who have done an outstanding job in crafting this manuscript into its final form.

To Molly Bergen and Katie Bergen for their unfailing enthusiasm and outstanding technical support.

And especially to my daughter Sue Bergen, who has constantly supported my efforts, edited and improved the text, and arranged and supervised every detail of its publication.

To all the above I offer my sincere thanks.

G.W. October 2018

Chapter One

1921–1931: The Early Years

As a little boy, I used to beg my father for a bedtime story, and he would often tell of his experiences during four long years with the King's Own Yorkshire Light Infantry (KOYLI, pronounced 'coyley') in the Flanders trenches. Many of his adventures were gruesome, such as his desperate efforts to scramble up the steep and slippery sides of bomb craters in an attempt to escape the stench of rotting corpses of men and animals. There were forays undertaken just before dawn into 'no man's land', crawling under barbed wire and through minefields to the edge of enemy trenches to listen for any signs of Germans massing for an attack. It's a wonder I ever got to sleep at all, but Father's voice droned on until I did. He frequently thanked God that he had emerged relatively unscathed, apart from the effects of mustard gas.

It must have been a wonderful thrill for him when, no longer 'Private Arthur J. Wilson', he returned to a peaceful England and met Nurse Edith Smith wandering round a quiet churchyard in Bottesford, near Melton Mowbray (Bottesford was an inconspicuous village until 1940, when it became the heart of an important Bomber Command Station). Both were filling in time, waiting for the next train to Leicester, Father's hometown and where Edith was working for the Institute for the Blind. Their subsequent married life in Bradford was not easy, and Mother had to physically restrain Father in the night, when he woke up yelling 'gas! gas!' and tried to jump out of the bedroom window. Nor was 1921 an auspicious year for me to be born, into a country struggling to recover from a disastrous World War, but I suppose I was lucky to have arrived at all, for Mother was forty when she produced me.

Father was born in 1884 in one of the poorest districts of Leicester, and as soon as he had a grasp of the three 'Rs' his mother decided that, now he was twelve years old, it was time for him to earn his own living. She whisked him off to the nearest factory where the manager took a liking to him, and offered him an apprenticeship leading to a good position. He was thrilled, but when his mother turned back and suggested 'The mill down the road might offer more money', the manager replied 'Better take him there then!' And that was the end of a wonderful opportunity.

Some experience as a bicycle salesman and four years in the Army did little to enhance his limited education, but everyone recognised his integrity and respected his determination to succeed. On her marriage, Mother's father immediately offered him a job in his garage called 'J S Smith and Son', but this didn't work out well, and he looked elsewhere.

At one time, Father acted as an appraiser for a firm of removal contractors, calculating the cost of a job and quoting a price for the contract. One day, he had an appointment with a woman to make an inspection, but when he arrived at the house no one seemed to be at home. The door was open, so he went in, made his appraisal, and left. Later the police came looking for him, as he had been seen leaving the house—with his ginger hair he was quite distinctive—and they told him that the lady had been upstairs all the time, having been murdered. He was not treated as a suspect, but he said that it had been 'an uncomfortable experience, Laddie'.

He was also employed as a guide for 'Mystery Tours'. For the vast majority of people who had no access to a motorcar, outings to the country in a *charabanc* (an expanded automobile for pleasure trips) were popular, and the secrecy about where on earth they were to be taken added spice to the trip. Bradford was fortunate in having a wealth of fascinating places within reach—Fountains Abbey, Brimham Rocks, the Ilkley moors, and historic castles at Skipton, Ripley, Bolton, Knaresborough and so on, all of which involved a journey through beautiful countryside.

The charabanc had at least three rows of seats seating four across, so the party would number perhaps a dozen. The driver wore a white uniform with a peaked cap, and I suppose Father's jodhpurs gave a sense of adventure to the event. He devised the route and commentary and enjoyed the work, for even from my earliest days he would never miss any opportunity to take me round any ancient monument—he made castles exciting!

Father and his 'Mystery Tour' charabanc.

Our first home was a modest house in Roslyn Terrace, identical to thousands of others built by mill owners in serried rows as close to their factories as possible. Every morning, we were woken up by a blast from the local factory claxon, accompanied by other hooters producing a cacophony that was not only awesome, but made domestic clocks quite unnecessary. It was a dreary environment. Everything was built of local millstone grit, with cast iron railings for safety, and as all the yards and footpaths were paved with stone, the only green things in our locality were weeds in the cracks of masonry and aspidistras in the front windows. Trees were rare, but I have to admit that Victorian efforts to provide sanitary housing, piped water, and gas for heating and lighting, together with proper drainage, made a great improvement on the appalling conditions that had prevailed earlier in the nineteenth century, when slum housing, over-crowding, open sewers and foul drinking water had reduced the average life expectancy in Bradford to just eighteen years!

Then Father found his first real job. He was invited to go into partnership with a Welshman, whose name I forget—let's call him 'Jones'. He was selling chemicals to mill owners troubled by hard scale that formed inside their boilers due to impurities in their water supplies. Jones evidently realised that Father's attractive personality was ideal for recruiting new customers, and things went well until some dissatisfied clients alerted Father to the fact that Jones's treatment was quite haphazard, being aimed at selling chemicals rather than actually solving problems. 'Permanent hardness' in the water required a different treatment from 'temporary hardness', and Jones's cavalier attitude didn't suit Father one bit, for his motto, which has stuck to me all my life was, 'If a job's worth doing, Laddie, it's worth doing well, or not at all'.

So the partnership was dissolved, and Father set up his own company incorporating in its title the words 'Scientific Service,' for he decided that the chemicals must be adjusted to suit each individual customer's needs. Samples of each mill's feed water were taken and then analysed by the best chemist he could find, Dr Lloyd, who was the head of the Chemistry Department at the Bradford Institute of Technology, now the University of Bradford. Dr Lloyd also gave advice on the mix and quantity of chemicals appropriate for each case, so that customers could be approached with confidence. The process of analysis impressed them, especially when follow-up treatments permitted costs to be reduced as the scale gradually diminished.

In theory, this was all very good. If Father had secured access to capital, his business would have thrived, but amidst the difficult economy of the 1920s, it took years of hard struggle to establish the little firm. To begin with, the split from Jones had been far from amicable. I remember, as a boy of four or five, being very frightened when I overheard Father speaking of Mr Jones's threatening retribution if Father were to approach any of his customers who, of course, were the biggest mills in the area. I imagined physical violence, whereas Jones probably envisaged no more than legal action. So, the search for customers had to go further afield, by train, tramcar, bus, and foot, since it would be years before Father could afford a car.

Chapter 1: 1921–1931 The Early Years

Father's first office was near the Bradford Alhambra, where Mother did all his early typing and filing, next to a small room where he mixed and bagged his powders. Success in his treatment of boilers demanded regular monitoring until the right mix was achieved. This could only be done when the boilers had been shut down for the weekend and had cooled off somewhat. Then, the boilermen could climb inside and chip off the hard scale with hammers, and Father would go with them. He said the noise was deafening, the heat intolerable, but his willingness to inspect the problems himself gained him the respect of the men, who assured him that none of his competitors would ever dream of climbing into a hot boiler. Trade was so bad during the Depression that I now wonder how his business survived at all. His industry, honesty, and integrity were such that, by the time the war broke out in 1939, he had his own car, excellent premises in Bolton Lane, a full time labourer to mix the powders, and two secretaries!

At the beginning, Father must have worked very long hours, for by 1925 he obtained a mortgage for £900—enough to buy a newly built semi-detached house in Undercliffe, one of the suburban communities based upon tiny villages that overlooked the City from the surrounding hills. I was only four at the time, but I remember the transformation of my life quite clearly. Gone were the candles and gas mantles, now replaced by the wonder of that age, 'electricity', that not only lit a single light bulb in every room, but also provided a primitive power outlet for those on the ground floor.

No other relatives had an indoor bathroom with a water closet like ours. It was the latest thing—better than any of the outside toilets used by the Smith relatives in Bradford, and far superior to the earth closets of the Wilson family in Leicester. It boasted a fireplace in the living room, and one in the front room, although the latter was only lit on Sundays, for coal fires were a luxury. The front room had a bay window with a window-seat, under which I kept my toys in boxes, carefully organised, so that they could be quickly packed away and shoved under the window seat if visitors appeared. My bedroom had space for a single bed, a dressing table, a stool, a wardrobe, and absolutely nothing else. There was an electric light fitting in the centre of the ceiling, and heating was provided by a ceramic hot water bottle in my bed, but compared with Roslyn Terrace, such accommodation was luxurious.

Unbelievably, we had two gardens, one at the back with a large vegetable patch, and one at the front with a lawn and a tree big enough for me to climb (when I first ventured to do so, mother was appalled at the filthy state of my clothes, for the tree was begrimed with years of soot). From the lawn, one could look right over the city to the hills on the far side. Of course, this view was usually obscured by smoke from the thousands of domestic coal fires, the only form of affordable heating, but on a hot summer day, especially during holidays when the mills were shut down, the panorama was breathtaking. One could make out the smallest details of buildings miles away, and even the time on the town hall clock.

The landscape was dominated by Lister's mill. It was originally built in 1838, and, following a fire, was rebuilt in 1871. Its colossal chimney was based on an Italian campanile. The top rim was said to be wide enough to allow a horse and cart to be driven round. It was the largest silk mill in Europe, employing 11,000 people and consuming 50,000 tons of coal each year (from its own private coal field). It was world famous for its silk, mohair plush, velvets, and imitation sealskin, and had supplied the silk curtains for the White House in Washington DC in 1896.

So, our move was a dramatic change, and to crown it all, at the bottom of our street was an entrance to Peel Park, one of the largest public parks in Bradford, built at the height of the City's Victorian prosperity. It had everything: a grand promenade with a huge flagpole, a bandstand, tennis courts, bowling greens, a café, lakes, waterfalls, and lots and lots of grass-covered open space in which I could play with my special friend, Malcolm, who was in the same class in the primary school. When Mother discovered public baths at the far end of the park, she insisted that Father and I should learn to swim (some years later I made my first 'rescue' when I fished out a little boy who had made the mistake of jumping in at the deep end. Malcolm helped me pull him out and the bath attendant got quite a lot of water out of him). I was useless on the high diving board, but later qualified for a 'rescue' badge, which proved to be far more useful than 'fancy' diving.

Mother and Father were ideally suited to each other, although they didn't always agree, for Father's slow but methodical approach to life tended to irritate Mother, who had far too much to do, and not enough time to do it, but he was loyal, loving, and honest, and would do anything within his power to help her. He might wield a paintbrush or write a good letter, but mechanical repairs, plumbing, and woodwork were quite beyond him.

His inability to do simple repairs must have come as a shock to Mother, whose father would tackle anything, having been trained as a blacksmith, and was now running a very successful garage where he serviced and repaired vehicles of every sort. He was known as an inventor, and had even installed locks in Armley Jail when he lived in Leeds. Mother had inherited her father's practical skills and tackled any problem at home with energetic enthusiasm, from hammering nails to the finest needlework.

Consequently, she expected me to make myself useful around the house from an early age. In the tradition of her father, she expected me to perform at least one good deed a day, exemplified by the idea of 'helping a blind man to cross the road'. Blind men were usually found in the city centre, where veterans blinded by mustard gas were not uncommon, but being skilled in the use of their white painted sticks and acute hearing attuned to traffic, they usually declined my offers of help. My main jobs were to light the living room fire before breakfast, and to help with the washing up. I was too small to reach into the kitchen sink, but I could dry the dishes and put them away. I learned about small coal (cheap and dusty) and large coal (expensive, but shiny clean) and how to screw newspaper into firelighters and chop up wooden boxes to make kindling for fires. By the age of ten, I was required to darn holes in my socks and to use Mother's cobbler's last to repair my shoes, and I found great satisfaction if I managed to please her.

One job I hated was weeding the lawn of dandelions and plantains, even though I was offered a reward of a penny a dozen. The only domestic chore I dreaded even more was the weekly wash; possing the blankets, sheets, and towels in a galvanised washtub and turning the handle of the huge mangle was backbreaking work. And it was very easy

Practicing one good deed a day.

to trap fingers in those rollers. Mother's posser consisted of a twelve-inch diameter copper cone, perforated with an elaborate pattern of holes, attached to a short pole. This was energetically forced up and down and round in the washtub to dislodge the dirt. It was hard work with a heavy load of sheets, and professional washerwomen developed such muscles as to be avoided in any pub brawl.

(Father's father, who was an innkeeper in the 1890s, once dashed out into the street to beg the local policeman to settle a violent dispute that was threatening to wreck his bar. He was explaining the situation to the 'bobby' as they approached the pub when the policeman suddenly stopped. 'Women?' He said, 'Did you say they were women? Oh no, Oliver, not if they're women!' And, with that, he walked away).

Later on, after the washing was done, one couldn't get anywhere near the fireplace because it was surrounded by clothing airing off after being hung on a line in the garden (a formidable job in winter, when things hanging on the line tended to freeze instead of dry). Then of course, everything had to be ironed—even the sheets—before being stacked away.

In addition to cooking, cleaning, sewing, knitting, and mending, Mother acted as secretary and bookkeeper for Father when he was setting up his own business. Even so, she made time to go to afternoon classes on cookery. As her journey home involved two tram rides at rush hour she would often be late. I remember feeling very forlorn as I sat at the top of the front steps waiting for her when I got home from infant school and found the door locked—but the goodies she brought home from the class always made up for it.

Where the name 'Undercliffe' came from, I never understood, for it was on high ground and one of many small villages that had occupied the hills surrounding the city from the Middle Ages. At that time, Bradford had been built on a 'broad ford' on the river Aire, to take advantage of the pure water emerging from the millstone grit that was perfect for scouring and dyeing wool from the sheep that ranged those hills. All the mills had been built along the valley bottom. There were none as high up as we were, so there were no ugly buildings to detract from the character of these small communities whose stone cottages clustered together in friendly informality.

My journey to school took me past a blacksmith's smithy (where horses were constantly being shod), an off-licence shop, (where alcohol was sold but not consumed, and which I was forbidden to enter) and our local fish and chip shop. Next, was the sweet shop, which supplied me with my weekly comic, 'The Rainbow', with the adventures of Tiger Tim. It was a pleasant walk, past generous open spaces, opening off the road where horses grazed, washing was hung out to dry, and where children could play in perfect safety, while mothers gossiped and watched from their open cottage doorways.

It was in one of these squares that I got into trouble, when sent to collect a paper from the newsagent. An itinerant peddler of crockery and kitchenware had set up his stall and was beguiling the locals with a patter designed to persuade even hardnosed Yorkshire folk to part with their cash. From the outside of the throng all I could hear was 'Only two shillin' and six—no—only two bob!' Pause. 'I tell you what—my very last offer—two for one and six!' I just had to see what he was selling, and it took some time for me to squeeze through the grown-ups to reach the front where I could see what was going on. I was spellbound, for by the end of his performance, it seemed to me that he was practically giving the stuff away.

Time for me just didn't exist, but for my parents it must have been an agonising wait, because the shop was only ten minutes' walk from our house. By the time I got home Mother was beside herself with worry and had given Father such a bad time that he actually threatened to hit me. But he was putting on an act to assuage Mother's feelings, for he would never actually hit anyone; he was the most gentle and loving man, fond of giving me nicknames such as 'Tommynod', 'Tom Noddin', 'Timothy Titus', or 'Sonny Jim'. Consequently, whenever he did give me a dressing-down, and called me 'a young varmint', I took him very seriously.

Parents accompanied their youngest children to the Infant School, but as they got older, and resented parental supervision, they were allowed to go in groups. So, this became a social event involving my friend Malcolm and Jack Pickles who lived in the grocer's shop at the top of our street. It was then, when nobody could overhear us, that gossip took place and mischief was plotted. In this tranquil environment, we were perfectly safe walking to school, so when on a tram ride to town, Father shocked me by pointing out some streets near the cathedral that I must never venture down. He said that even police went in pairs for their own safety. They were a good mile away from home, and Mother confidently sent me on local errands to Barnett's the grocers' or to the butcher's to buy eggs. I hated this last chore—the eggs were counted into a paper bag and rubbed against each other until one or more of them broke and I had the messy job of washing the sticky yolks and whites off all the survivors. Mother finally solved this problem by providing a basket with crumpled newspaper packing at the bottom (egg trays patented by Joseph Coyle in Canada in 1918 had not yet reached the UK).

Chapter 1: 1921–1931 The Early Years

The village was simply a crossroads, with one public house and a surprising number of shops clustered round about, all of which were incredibly tiny by modern standards. A tobacconist had space for himself and one customer only; the cobbler had room for two customers. Other shops were a bit larger, all owned by people who lived on the premises, knew their customers by name, and who loved to gossip. Consequently, shopping tended to be a social activity and quite pleasurable for those with time to spare. The most prominent shop, at the tram terminus, was 'Sparkes', a baker and confectioner, whose large glass window must have been at least five feet wide. This always displayed a mouth-watering selection of cream cakes and tarts. It made the much smaller shops nearby look tawdry and mean—especially the pharmacist, whose dreary window displayed nothing but huge bottles filled with coloured water. But my visits to Sparkes were rare indeed, for Mother baked her own bread and had learned how to make excellent cakes and pastries.

Housewives had to be pretty fit to climb the hill and trudge from shop to shop, usually queuing at each one with bags getting heavier at each stop, and then return home carrying all the shopping. Saturday visits to Bradford were even more strenuous, with shops much further apart and steeper hills. Mother was thankful to have a small boy to help on these outings as she often went to Lingards—the pre-eminent drapers and household furnishers—for 'bits and pieces', for this was the 1930s when the practice of 'make-do and mend' dominated most households. This emporium had the latest mechanical invention, designed to avoid unskilled workers having to handle money and deal with accounts. All the shop assistant had to do was make out a bill and put it with the customer's money in a little basket suspended from a cable, and then pull a lever. This catapulted the basket at high speed along the cable to a central kiosk, high above the concourse, where the supervisor and his clerks could look down upon everything in the shop and deal with the cash and accounts. I loved to watch the baskets whizzing to and fro along the wires, and longed to have a go with the lever to see if I could make them go any faster, but it was all well out of my reach. All I could do was to wait, impatiently, for its return with the receipt and change, while the assistant chatted with my mother and tried to interest her in another purchase. Some hope!

Like most Bradfordians, I entertained acerbic views about competition from shops in Leeds, and when I discovered a departmental store using cylinders that were sucked through a vacuum tube with a loud 'Swoosh', I refused to admit that this idea might be more efficient than Lingards' elegant flying trapeze and likened it to a sewerage system. I dismissed it as typical of the sort of thing Leeds folk *would* like. However, the Germans resolved the issue in August 1940 by dropping a stick of bombs that destroyed the whole of Lingards' store, and it was never rebuilt.

I suppose, because his father had made shoes, Father was extraordinarily particular about them, polishing them daily, and saying that people were judged by their footwear. He was proved right some thirty years later, when my wife and I were in an over-crowded bookshop in Leeds. She turned to me and whispered, 'Isn't that Sir Roger Stevens in front of us?' When I asked how she knew it was the new Vice-Chancellor she said, 'It was easy—I could see his shoes!'

My school shoes were not of his quality and were constantly in need of resoling and heeling. I remember the warning signs vividly. A nail would project through the inner sole and make a sore place on my foot, and I had to insert a piece of cardboard until I got home. Repairers sometimes took several days; the small cobbler was cheaper, but so busy that we were sometimes driven to use the bigger establishment that sold saddles and all types of leather goods, such as handbags, school satchels, brief cases, and so on. This shop was patronised by the local 'gentry' who wouldn't be seen going near our little cobbler's, but even there, one couldn't buy a new pair of shoes. That was the province of a separate shoe shop that did a roaring trade in children's footwear, which wore out even faster than that of the grown-ups. It was not surprising that Mother bought her cobbler's last that continued to serve me well for the next seventy years. In this tiny community, each shopkeeper was a specialist. The 'tool shop' provided anything from a wheelbarrow to hammer and nails, but to find pots and pans or a yard brush, one had to go to a 'hardware' shop, frequented mainly by housewives, and hard feelings arose if shopkeepers impinged on each other's territory.

The pork butcher killed his own pigs and made black pudding (a northern speciality made from pig's blood), and the butcher had his

Chapter 1: 1921–1931 The Early Years

private supply of seasonal game, always reserved for special customers. Mother's favourite purchases were rabbit for her rabbit pie, a (more expensive) chicken for Sunday lunch if we were having visitors, or if it was a VERY special occasion, a piece of salmon, which required a journey down town to the only source of fresh salmon, the fishmonger in Kirkgate Market.

On the other hand, there was plenty of cheap cod around—but only in the fish and chip shops. There were five of these within walking distance of our house, with opening hours coordinated so that they didn't conflict with each other. The only trouble was in remembering which one was open, and at what times. In this I soon became an expert, finding out the hard way—by making fruitless journeys. They all provided a generous portion of cod and a pennyworth of chips for three pence ha'penny. Considering that a Mars Bar cost two pence, this was a very reasonable price to pay for what was the main meal of the day for many households. They were sold on a strictly 'carry-out' basis until Harry Ramsden opened his first fish and chip restaurant in 1931 with tables and chairs and a pot of tea—at an additional charge of course. He had deliveries of fresh fish sent from Grimsby every day, and his business never faltered. He was a clever lad—the restaurant was sited at Guiseley—half way between Bradford and most of the beauty spots in Wharfedale, and not far from the newly opened Leeds and Bradford airport at Yeadon, so it was a natural oasis for families on their way home after a day's outing, now that the 'upper-middle class' could afford a motor car. He even provided a car park.

A popular make of motor car was a Jowett, manufactured in Idle, an adjacent village that lay at the end of Idle road, which passed the top of our street (made notorious by music hall jokes about its Idle Working Men's Club). The firm had pioneered engines from 1899 and produced their first complete motorcar in 1904. The 1932 models were modest and rather frail looking, but were reliable and economical, having a seven-horse power engine with only two cylinders. Later versions were more sophisticated and good enough for Scotland Yard to buy for their Flying Squad. Vintage Jowett Javelins are now prized, especially in Australia and the USA.

The only substantial building in Undercliffe, apart from the church and the chapel, was the cinema, which unkind people called the 'flea-pit', for it was very small and only showed well-worn silent films. Mother rarely let me go there, for she 'didn't know how often it was cleaned'. The entrance fee was one penny, and the absence of any 'sound', apart from the feeble piano, was made up for by exuberant shouting and cat-calls, especially during cowboy films, which were shown mostly in serial form, so that if one missed an episode there was real grief in the family.

Mother and Father knew that I would willingly forego this pleasure if there was any chance of going with them to one of the magnificent new 'Picture Palaces' being built in the centre of Bradford. Their programmes lasted three or four hours, with a 'supporting' film, a newsreel, an intermission for the sale of ice cream that was brought to one's seat by a young lady, followed by the main feature. Talk about luxury! And some films were 'talkies' too!

The only thing good about the intermission was the ice cream. I thought nothing of the new 'Wurlitzer' electric organs that rose dramatically from the orchestra pit as the lights went up. They produced the most disagreeable noises, particularly when attempting to imitate instruments such as drums, trumpets, or even a xylophone. Yet people loved them and applauded the organist, especially if he happened to be Reginald Dixon of the Tower Ballroom, Blackpool, who regularly gave organ recitals on BBC Radio.

Refreshments enhanced these Saturday afternoon jaunts into town. Father was very fond of sweets and you can imagine my delight when he brought home a seven-pound jar of Needler's assorted éclairs and announced that one of his clients had let him buy sweets at wholesale prices. Until that happy day, we relied entirely on a cut-price sweetshop in town, and to complete the outing in style we might slip into the little café just outside the cinema for afternoon tea after the show. If Mother couldn't afford this extravagance, she would climb all the way up the hill to a dairy outside Kirkgate Market, where she would buy a carton of thick cream or some cream cakes to take home.

There was no dairy anywhere near home. Milk was delivered to our doorstep every day by a milkman who had cows on a nearby farm and

Chapter 1: 1921–1931 The Early Years

who ladled the milk directly from his churn into our own jugs. The only way we could get cream was by leaving the milk overnight in a wide bowl covered with cheesecloth to keep off the flies and skim the cream off next morning (or surreptitiously dip in your finger and lick it when no-one was watching). In summer, the only way to avoid milk going sour was to immerse the jug in a bowl of cold water or boil it as soon as it was delivered. In case of emergencies, Mother always had a tin of Nestlé's condensed milk in reserve. Father disliked it intensely. In his youth, he had been caught in the larder sampling this sickly-sweet milk from the tin and had been kept there until he had consumed the whole tin. It put him off the stuff forever (this wasn't the first time he had been in trouble—his father once found him desperately gasping for breath after sampling some high-proof gin that was waiting to be diluted before being sold to his customers in his pub).

Thank goodness he still liked ice cream. On one memorable day, we walked together over Ilkley Moor and emerged near the Cow and Calf rocks to discover a 'stop me and buy one' ice cream tricycle. It had been very hot over the moor, so Father bought us each a tub. I was ready to resume our walk when he dumbfounded me by asking 'Would you like another one?' I thought he must be joking, for the Wilson motto was 'Moderation in all things' (on Mother's part it was more like 'Extreme moderation in all things'), but he was serious and he finished his second tub with relish. I was in for a shock. It was incredible—I couldn't finish my second tub! I would never have believed that there was a limit to the amount of ice cream that I could put away. This turned out to be the first of many sad lessons about how unfair life could be.

Of course, Mother's two sisters had already found this out, when the War deprived them of any chance of finding husbands, even though they were attractive and intelligent. Being the oldest, Aunt Anne assumed duties as housekeeper to Grandfather and his son, Arnold, and Aunt Mabel found herself nursemaid and companion in Doctor Crawford's family in Manningham Lane. Edith was the prettiest of the three, and I suspect she attracted a certain amount of jealousy in having secured a husband and having a son whom they liked to entertain.

So when Mother could be persuaded to let me go, they would collect me from Sunday School in Morley Street and we would walk to

Grandfather's—a distance of about two miles, which seemed a very long way for a little boy of three or four. A tramcar could have taken us partway, but they said the walk would do us all good—and it was cheaper. It was a very boring route, and I liked the graffiti on a railway hoarding that said, 'ITS A MEAN OLD SCENE' (and was delighted after the war to find it still intact). Boredom was further relieved when we passed a haberdasher's shop owned by a Mr Longbottom. This name invariably sent me into paroxysms of laughter, which were prolonged by references to Messrs Sidebottom, Winterbottom, and even a Mrs Shufflebottom, so that I usually arrived at Arncliffe Terrace giggling and in high spirits.

We entered from the back yard, by the midden (where the refuse bins were), and thence past Uncle Arnold's aviary of budgerigars to enter the kitchen-cum-living room. Polly the parrot dominated this room from a cage hanging at eye level immediately next to the door and she squawked loudly at anyone who came in—an alarming experience for unwary strangers. She had been very slow to start talking; Aunt Mabel had spent weeks patiently repeating words such as 'Hello!' and 'Good morning!' without the slightest result and had despaired of getting the bird to speak, until one day, she came in loaded with Christmas shopping and a strident voice in her

Together with Grandad and Polly

ear squawked 'MERRY CHRISTMAS!' She dropped her parcels in alarm and called out, 'Who's there?', but of course it was Polly, who never tired of talking after that. Aunt Mabel had no difficulty whatsoever in teaching her nursery rhymes, monologues, and even songs. She repeated odd phrases and remarks uttered by members of the household, but when she started to include 'Damn and blast!'. Grandfather, a pillar of the local Baptist Church and a Scoutmaster, had to tell Uncle Arnold to watch his language.

Sunday dinners at Grandfather's were the most important meal of the week and invariably consisted of cold brisket with onions, followed by apple pie and custard. The onions I hated but dared not say so. Not that I was averse to experimentation; when a little bowl containing a smooth yellow mixture was placed near me, I took the delightful little spoon especially designed for it into my mouth and savoured it. There was a stunned silence as all the grown-ups waited for a howl of distress, for this was fiery English mustard—hence the need for the miniscule spoon! But I liked it and have never since tired of mustard in any form. When I got home, I was honoured by being given the responsibility of looking after the mustard jar—mixing fresh mustard from powder, milk and water whenever needed, and making sure the edge of the jar did not get crusty—which it did very quickly as the mustard dried. It was my special job.

Uncle Arnold enhanced the prestige of the garage by becoming very successful at motorcycle rallies. He had a phenomenal collection of trophies and challenge cups in display cases and on the mantelpiece. He bred budgerigars, and when one perched on the neb of my school cap and peered upside down into my face, Uncle said I could keep it. Paddy proved to be the dearest, sweetest pet you could imagine, and was surprisingly intelligent—he could not only repeat phrases and rhymes, but he could imitate Father's whistling. When it was nearing lunchtime, he would anticipate Father's arrival by flying to and from the front door with a great deal of fluttering and squawking. He slept in his cage at night, but otherwise had the freedom of the house. He flew onto my hand or shoulder the moment I came into the room. When I talked to him he would close his eyes and rub his beak gently against my nose, and if he was on my shoulder he had a delightful habit

Uncle Arnold with some of his trophies

of delicately nibbling at my neck or ear in affection. Paddy's favourite perch was on the bottom frame of a large mirror, where he would chatter and argue with his reflection until he was exhausted. When Mother was sewing, he would steal buttons, carry them to the edge of the table and drop them to make a pile on the floor, which he would visit from time-to-time chortling, as if to count up his loot.

We usually spent evenings reading by the fire. Paddy would perch on my book with his eyes shut, one claw on each side of the spine, holding the pages down. When I wanted to turn a page he would lift his left claw and then the right claw to hold down the newly turned page. It was an awkward procedure, and after a few pages he would tire of it and fly off, but the following evening he would insist on repeating the performance. To liven things up in the evening, he invented a game. We kept matches for lighting the fire in a metal box on top of the mantelpiece. He would get his beak on to the edge of the box and shove it along the full length of the shelf—it was difficult because his claws tended to slip on the tiled surface. He would struggle along to the end, and the metal box would make a tremendous clatter as it

Chapter 1: 1921–1931 The Early Years

plunged onto the tiles below. He would swoop down with a chortle of delight to examine the damage. Alas! One day, Mother had to open the front door to an unexpected visitor who kept her talking, and Paddy flew out. It was during a snowstorm, and although we whistled and searched outside for a long time, we never saw him again.

It was 1925 before Father could buy a radio set, for there had been no electricity in Roslyn Terrace, and I was not very popular when I accidentally knocked it and put it out of action. No one thought to tell me how delicate it was, with about six thermionic valves with filaments that were fragile when red hot. Father had to take it into Bradford to get it repaired, but Uncle Arnold showed me how to test bulbs to find the faulty one, avoiding any more journeys by tramcar into town hulking a bulky radio set. It was he who gave me my first radio, comprising a pair of headphones connected to a box with a fine wire, known as a 'cat's whisker', which I had to manoeuvre so that it JUST touched a lump of crystal. It didn't have a battery, but it somehow received the BBC signals. I liked my little crystal set, which I could secrete under my pillow and listen to at night. It took a great deal of patience to find a spot that would produce a result, but as '2LO' was the only broadcasting station, tuning didn't come into it. In practice, the slightest movement broke the critical contact with the crystal, and then I had to start all over again. I soon fell asleep.

In those days, all respectable families had to have a piano. It was a status symbol; the equivalent of today's television set. Pianos were everywhere I went, so I had to learn, and I had a dreary teacher for many months. When I learned an exciting tune

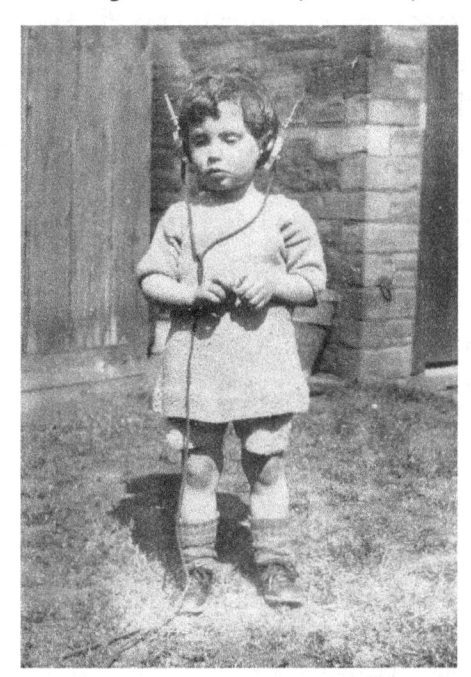

The author listening to 2LO circa 1925—note the home-made clothes!

such as Gypsy Rondo, I took pride in perfecting it. But, sight-reading defeated me, and it wasn't until sixty-five years later, when I took an interest in my granddaughter's lessons, that it dawned on me that 'middle C' was half way between the upper and lower lines of music staves. Of course! That made sense—if only the stupid man had told me that, I might have enjoyed learning to sight read.

Grandfather had been inspired by Baden-Powell's visit to Bradford in 1914 to found a Boy Scout Troop at the Tetley Street Baptist Chapel. I shall never forget the time when he introduced me to his scouts one Sunday and whispered, 'You'll be the Troop Mascot!' and took me by the hand in front of the band as we marched up to Lidgett Green. I tried to keep in step with his much longer stride as he kept in step with the band.

In front of us was the Troop Leader, carrying the Union Jack, and a Patrol Leader with the Troop Colours and two escorts. Behind us, the kettledrums rattled, the big drum thumped, and I remember the thrill that ran up my spine when Grandfather raised his hand and the bugles blared forth. I must say it made an impression on me, for even now, ninety years later, I can still recall the tune. Of course it was a very simple one, for a bugle is a primitive instrument with no valves; only the most skilful players can get more than four notes out of it, and not many of the boys had that capacity. Indeed, the band practice in the chapel is an experience my ears will never forget! But, in spite of some discordance, they managed to produce a very stirring march, enough to make people stop and watch us go by. It seemed to be an awfully long way, but I suppose it could not have been more than two miles before we arrived back for the Sunday Service when the flags were reverently laid before the altar and Grandfather read from the Bible.

It was in 1929 that Grandfather took me to the Scout Jamboree at Arrowe Park in Birkenhead, near Liverpool, to hear the Chief Scout, Lord Baden-Powell, speak. Although I was a bit too young to really appreciate the experience, I have never forgotten it (it was at Birkenhead that the first two Scout troops in the world were formed in 1908, and the second Birkenhead troop is still going strong—the oldest troop in the world).

Grandfather delighted me by announcing that he was going to take the family on a picnic. There was only one snag; he could not bring his

Chapter 1: 1921–1931 The Early Years

Grandad and his scouts go camping

car down to our house because Valley View Grove was just as the builders had left it, with huge potholes and protruding stones that would wreck his car before we started. Of course, Malcolm and I loved a playground such as this, where we built sandcastles and made racetracks for our Dinky racing cars, and it was free of all traffic.

So, we duly carried everything to the top of the street where Grandfather waited for us. This was to be my very first ride in a motorcar; I remember that it had a running board along the side because I barked my shins when clambering up. Being an open tourer, we had to wrap ourselves in travelling rugs and scarves to keep warm, even though it was a sunny day. I remember seeing aeroplanes at Yeadon aerodrome, (they must have been Avro 504K trainers used by the RAF during WWI, which regularly flew over our house when showing sight-seers over Bradford). We were then driving into the countryside for a while, until we came to a very big hill with a cart track winding its way to the top. As we started to climb, we could see several vehicles in distress further up the hill, which was perilously steep in places. At one point we had to pass a car with steam spewing from its bonnet. Manoeuvring near the edge, overlooking the steep slope, became so tricky that, for safety's sake and much to my disappointment, the passengers had to get out. I found it all terribly exciting, especially when Grandfather said that the climb was renowned as a severe test of man and vehicle, and on average more than half the cars failed to reach the top. Of course, there were no mobile phones or radios in those days, and drivers had to do their own repairs on the spot. Every car had at least two spare tyres, because tarmacadam was unknown in the countryside, and punctures were very frequent. Grandfather carried four spare tyres, and repaired a couple of cars on the hill, which I suppose did his garage business some good.

Several miles further on, we came to a famous 'beauty spot' with enormous rocks, where we had our picnic lunch. On the journey home, we had another impromptu picnic, in a tree-lined avenue, which I now realise must have been the drive leading up to Harewood House, while Grandfather attended to a puncture and lit the acetylene lamps to guide us home.

(It must have been more than thirty years later that I was on my regular commute from university to our home at Woolwich Farm and climbing Norwood Edge for the umpteenth time, when I suddenly realised that this must have been *that* hill! In the years between, the road was transformed, completely rebuilt, and re-surfaced, and while we were living in Woolwich, it was further improved with a smoother gradient and transition curves. Even with these improvements, it was still the scene of many accidents, including a memorable one when an oil tanker failed to take a bend and rolled over and over down a steep field. Later still, I realised that it was Brimham Rocks, where we regularly exercised our dogs that had been the setting for that long-ago picnic. These moments of revelation gave me a curious, indescribable thrill of déjà vu; it was like being reborn).

The memorable picnic at Brimham Rocks was to be eclipsed by an even bigger adventure when Uncle Cecil Wilkinson persuaded my parents to let me spend a whole weekend with him and Aunt Gertrude on a trip on his new motorcycle to North Wales. I was to ride in the side-car, and Aunt Gertrude would ride pillion behind him!

It was raining heavily, and the sidecar gave me some, but not a lot of, protection. We had got as far as Canal Road in Shipley, three or four miles away, when we heard a policeman furiously blowing his whistle. Cecil dutifully stopped and found that the local traffic cop was pursuing him on his pedal bicycle. I was petrified—what had Uncle done to get arrested? Were we going to spend the weekend in jail? It turned out that the red light that should have been showing at the rear of the bike was not lit. We were let off with a warning to get it fixed at the very

Chapter 1: 1921–1931 The Early Years

Uncle Cecil was proud of his new motorcycle.

next garage. As these were few and far between in those days, I spent a very worrying time until we eventually found one open, together with a very welcome bed and breakfast place with rooms for three very wet travellers.

My main recollection of that weekend was of seeing a mountain on fire. I found this very exciting because there was lots of smoke and flames and I couldn't wait to tell Mother and Father about it, for I was sure it was a volcano about to erupt. It was very disappointing to be told that it was not the mountain, but only the heather on it that was burning, and everyone expected me to be pleased that the rain, which continued to pour down, was expected to put the fire out! I was very glad to get home.

I became very fond of Uncle Cecil and Aunt Gertrude. She was an elementary school teacher and the only member of my family who was well educated and so could be regarded as 'cultured'. Cecil was not religious, but he was an idealist, and greatly admired the efforts of Sir Titus Salt to create an 'ideal village' for the workers in his new mill, that he had built by the river Aire, in open country well away from the grim environment of Bradford. He built good housing with plenty of open space and light, but without any licensed premises, for the sale of alcohol was, and still is, forbidden in Saltaire.

Gertrude and Cecil chose to be married (with me as page-boy), in the Saltaire Church, which is now highly regarded by architectural historians. The whole of Saltaire village is now a World Heritage Centre, with the mill complex converted into a cultural centre with a theatre and art gallery and is now a major attraction in West Yorkshire.

Cecil had a responsible position as the buyer of wool for a large firm and had friends and colleagues of 'substance and influence' who played golf and bridge. Visits to stay with Aunt Gertrude were enjoyable, but especially for the gramophone (wind-up, of course) that played music such as I had never heard before. It is true that Father played the piano, but only hymns, which he thumped out laboriously, so this was the first time I had heard orchestral music. I was thrilled by the excitement and beauty of what I think must have been classical ballet, and I danced and pranced about until I fell breathless onto a huge pile of cushions next to my aunt. I realise now that she and Cecil must have made a special effort to introduce me to 'culture' as distinct from 'education', although I'm afraid that most of the expensive books that they always gave me for Christmas and birthday presents (including an Encyclopaedia of Butterflies and Moths) were soon put aside in favour of more normal toys.

Both were passionately keen on the theatre and were founder members of the first Civic Playhouse in Bradford, where Aunt Mabel was to act in several plays. She was the only one of the Smith family who really got on with the Wilkinsons, who really appreciated her talents, whereas she always seemed to be treated by Anne and Edith as a rebel. Actually, she was the sweetest person, whose only rebellion was against the interminable religious and political arguments to which she had to listen all too often, and I often wonder whether the Victorian prejudice against anyone who acted on the stage might be to blame for their attitude towards her.

When Aunt Gertrude arranged for me to attend a 'Young Actor's class' at the Playhouse, I found the proceedings incomprehensible and evidently designed for people enthusiastic about going on the stage, which I was too horrified to even contemplate. I was inside the play-

Chapter 1: 1921–1931 The Early Years

house giving in my notice to quit when some villain stole the pump from my beloved new bicycle that I had carefully chained to the lamp-post in front of the theatre. That really settled it—no more acting for me! The police were called to deal with this heinous crime, and to my amazement they actually caught the thief. He was brought to juvenile court, where I had to give evidence, and the pump was restored to me.

Many years later, at the Civic Playhouse, I saw my first play by J B Priestley, who had attended Belle Vue School, as I was to do in the future. He was revered in Bradford, and especially by Cecil and Gertrude, because he was a fervent socialist and the most famous personality that Bradford had produced so far. His weekly broadcasts during 1939 and 1940 were second in popularity only to those given by Churchill. They were called 'fireside chats' and his deep voice was so relaxed that it gave comfortable reassurance to listeners undergoing a whole series of unnerving and worrying wartime crises. For, at that time, the Germans were bombing Britain and sweeping all before them on the continent and Northern Africa. But, believe it or not, Churchill was actually persuaded to intervene and stop Priestley's talks because members of his cabinet found them too 'left wing' for their comfort.

Uncle Cecil's Christmas parties for the family were embellished by gifts from firms wishing to cultivate his patronage—wines, spirits, chocolates and cigars. Mother, Anne, and Mabel, always came home complaining that they 'would never get rid of the smell of cigars and cigarettes in their clothes.' But I liked the smell of cigar smoke, and also the chocolates (especially those with liqueurs inside them) and I really didn't mind if Uncle and Aunt 'smoked like chimneys'.

What sometimes spoiled these events for me were the heated arguments that inevitably arose about religion and politics, which at my age I didn't understand. After the loss of so many young men in the First World War, beliefs in spiritualism and faith-healing had a big following. Cecil was very sceptical about those pursuits, and never lost an opportunity to criticise the Christian Science beliefs that Mother and her sisters held so fervently.

Mother was a resolute Conservative, whereas Cecil's father, Uncle Tom Wilkinson, who I remember as a small, very old man, who smelled of stale cigarette smoke and rarely spoke to me, had been a

City Councillor in Bradford for many years representing the Labour Party. Mother regarded their politics as perilously close to Communism. Cecil vigorously endorsed his father's opinions, and as this was a time when the Soviets were coming to power in Russia and threatening the rest of Europe, Mother was understandably worried. It was perhaps just as well that the three successive tram rides that brought us home to Bradford were so tedious and noisy that the wrath of my agitated mother had time to cool off before she put me to bed.

In general, I was very happy, but in sharp contrast to this idyllic lifestyle, I was faced with the grim realities of life by spending every alternate Christmas in Leicester, 'the worst slum in England', where Father had been brought up, and his sisters still lived. Mother found this a strain, but, with a brave face, allowed me to taste Aunt Emma's potato wine, saying 'being home-made it must be alright'—in fact it was extremely potent, but served to ease her tension. Aunt Emma lived in perpetual gloom; she was always swathed in a heavy black dress with at least five black petticoats, and her flowing skirt trailed on the non-too-clean floor. Very little daylight penetrated her living room, festooned with dark curtains and draperies, and the single lamp was only lit if we started to play cards, for any artificial light was 'expensive'.

She used to have a little corner shop and told us of one regular customer who would come in just before the lamp was lit and would pay her bill with what looked like a gold sovereign, but which was actually a highly burnished farthing. As the bill was rarely more than a shilling or two, she not only got away with the goods, but a sizeable quantity of change as well! Aunt Emma said that even when she was found out, there was no means of telling how long this had been happening.

Two other sisters, Eliza and Pamela had a bedroom for Mother and Father in their house a few streets away, and a pokey little garret for me above the outdoor scullery, with room for a small bed and nothing else. Everybody had stoneware hot water bottles for the coldest weather. There was no bathroom, of course, and the only toilet was at the bottom of the yard, with pieces of newspaper for toilet paper.

Cooking was done in the living room on a huge range, with its fireplace heating the room, the oven, and cooking pots. With the dining table in the middle of the room, there was no spare space, but an upright piano *had* to be crammed in somehow. When five people sat down at the table for a meal (or any other activity) two had to squeeze between the piano and the table, and two between the table and the sideboard, leaving just enough room for the fifth to serve the meal. The front room was even smaller, and was remembered mainly for its horrible horsehair settee, a punishment on my bare legs. The only other furniture was a chair, and a table in the window supporting an aspidistra, as a sign of respectability. We never used it, for it was reserved for special events such as funerals.

Aunt Eliza's acerbic manners frightened me (she never smiled), but Aunt Pam was plump and jolly, and made evenings with card games and chocolate raisins pass very pleasantly, especially if Father's sister-in-law, Aunt Polly, joined us, for both of them delighted in passing candies to me even when Mother said I had had enough.

The journeys to Leicester were stressful; railways were slow and unreliable, and the Great North Road was still a two-lane highway on which it was difficult to pass heavy trucks in snow and fog. Father was very pleased if he could average twenty-five miles an hour over the 200-mile journey, lasting eight or nine tedious hours. So, visits to Leicester were a burden and I always looked eagerly forward to the next Christmas, which I *knew* would be celebrated in Bradford.

By 1930, life was becoming a happy experience; Malcolm Shaw (known to me as Mac) was my best friend, although his parents were better off than mine. He introduced me to flying; my little box kite flew very well, but Mac's kite was much bigger and flew higher. We each had an assortment of gliders before we graduated to FROG (acronym for 'Flies Right Off the Ground') models, which had an ingenious gearbox that made the process of winding up the elastic band that drove the airscrew easy. Of course, Mac's plane was bigger than mine and flew higher and further (a model FROG aeroplane, exactly like mine, was recently shown on the Antiques Roadshow as a very collectable item). Mine disappeared with all my other toys that were given to a charity during the war when such things were no longer available. We sailed our model yachts on the lake in Peel Park (need I compare their size?) and we once took Mac's father's schooner to Yeadon, where there was a dam near the aerodrome large enough for such a splendid model. The schooner was well beyond our competence to handle properly, but it was exciting to try.

In the realm of winter sports, we were more evenly balanced; Mac's toboggan had broad runner blades that flattened fresh snow nicely, even on the grassy slopes of the Park, whereas mine had narrow blades that dug into the grass and got stuck. But once a track had been formed and frozen overnight, my sled would go much faster than his because the friction was less. Indeed, down the very steep hill next to the fish and chip shop, with its cobbled stone surface smooth with ice, mine flew downhill at breakneck speed. Unfortunately, the problem of how to stop it became serious, because there was no 'run-out' at the bottom— merely a row of bent and twisted iron railings. These were intended to prevent people from falling into the allotment gardens, some ten feet below, and a good many of the railings were missing. It was certainly exciting—not to say foolhardy—and after much experimenting and near disasters, we devised a method of slewing the sledge round and rolling off it at the last minute. It was an uncomfortable manoeuvre, and sometimes painful; I think we were secretly relieved when some of the parents living in the street realised what we were doing and put a stop to it.

I was very happy at the Undercliffe Primary School, which encouraged handicrafts. My very first model was a tramcar, fashioned crudely from the two halves of a box that had held North African dates. This effort was so highly praised that I graduated to carving model ships, such as the Santa Maria out of balsa wood. The rigging and ratlines had to be fashioned from fine cotton, which, on a model that was only four inches long, was not easy. They culminated in the slightly larger Bounty, which had a tiny brass cannon, and this survived, with some minor battle damage until the 1960s, when it was accidentally swept off the mantelpiece with catastrophic results.

Mac became my lifelong friend; he had a most attractive personality. Indeed, he was so sweet natured that I never felt any jealousy over his more affluent background and lifestyle. We became separated during the war. He was in the RAF, while I was posted to London, but we renewed our friendship after the war and corresponded until he died, well into his eighties. It was only then that I discovered that he had named his son, whom I never met, 'Geoffrey'.

Chapter Two

1931–1938: High School and Scouting

I was ten years old when my teacher arranged for me to take the entrance exam for the Bradford Grammar School. English and arithmetic tests were similar to those I was accustomed to, but there was another section, presenting peculiar shapes, diagrams, and puzzles, which baffled me at first, and by the time I realised what was required, I had no time left to work out the answers. So I failed. However, my results from the standard 'Eleven plus' exam were good enough for Belle Vue High School, and I went there happily enough with Malcolm—only to come a cropper in my English Literature exam.

I thought I had done brilliantly, for I wrote page after page with great enthusiasm for the whole time allowed. Mr Bradshaw explained that I had failed because I was expected to answer six questions, not just one! These two experiences stimulated an expertise in examination techniques, which became invaluable in later years; I never went in for any examination without carefully analysing previous papers and having a stopwatch on my desk during the exam itself.

The journey to school now involved a tram ride to Forster Square, and then another along Manningham Lane. As the tramcars were slow and noisy, and ran on an uncertain timetable, they really could not be trusted to get a worried schoolboy to school on time. The Bradford trams were particularly hideous; their axles and wheels were rigidly fixed to the chassis so that variations from a straight line were only achieved by the cast-iron wheels being forced round by the curved rails with much shrieking and grinding. As the track got older and badly worn, the rails sank into the cobbled road unevenly so that, even on the straight and level, trams would lurch drunkenly from side to side. Bradford was surrounded by seven steep hills, and turning a corner going downhill could be a hair-raising experience. I remember Aunt Polly, on a visit from Leicester, being so terrified that Father had to reassure her by explaining that the tram could not fall over 'because it was held up by the pole attached to the overhead wire'. I stared at Father in disbelief; surely he did not really believe that? I opened my mouth intending to protest at such a blatant untruth when he gave me a very stern look and winked at me. This was my first experience of a grown-up white lie.

I was about eleven years old. Malcolm and I were on our way to school when the tram reached the top of Church Bank—a steep hill next to the cathedral—and we began to clamber down the staircase ready to jump off to change trams in Forster Square. As we reached the bottom of the hill, the driver slammed on the brakes and we nearly lost our grip on the handrail as we were thrown sharply forward. The man in front of us wasn't so lucky; he lost his grip and hit his head on the metal framework with a crack.

'Damn!' he said, 'Damn, damn, double damn, two hells and a bugger!' Malcolm and I were spellbound with admiration at such eloquent and mellifluous swearing. On repeating the phrase it seemed to acquire a poetic cadence—indeed almost tuneful—and we practised it whenever possible until Father overheard me one day and forbade its future use. Of course we had no idea what the last word meant, and he did not enlighten us. It remained a mystery to me for several years.

Then, for my twelfth birthday, Aunt Emma sent Father enough money to buy me a proper bicycle. I was in heaven and was delighted with the City Council for making Valley View Grove part of the new Ring Road, which now had a beautiful asphalt surface that swept past our front door. I could now get to school in less than half an hour, and even come home for lunch. Although I was proud of my Raleigh bike, I have to admit that it was the cheapest available. Malcolm's had cost five *guineas* (five shillings more than mine) and it had a three-speed gear and metal mudguards compared to my Bakelite mudguards that were always breaking. So, spending money was devoted to upgrading my precious vehicle. My brittle mudguards were replaced by flexible celluloid ones, and a Sturmey Archer three-speed gear was installed in the back wheel (actually this was a necessity to tackle the steep hills to and from school.)

By this time, Malcolm had added a derailleur three-speed gear, to make nine gears in all, but he admitted (under terms of strict secrecy) that changing gears was so complicated that he wished he had never heard of a derailleur. My first bicycle lamp burned paraffin; it was dirty and smelly and was succeeded by a new lamp with a battery that was 'dry' until it leaked acid and eroded the metal casing into a horrible sticky mess that meant buying a new lamp.

Much later, I invested in a dynamo. This was fixed to the frame and could be flipped to touch the rear tyre to serve the front headlight and a rear red lamp. This was the last word in sophistication, and what a boon during the five war years when batteries were very difficult to find!

On their very first day, all new boys at Belle Vue were warned that they must not be seen speaking to anyone in the girl's school, which was in the middle of one large block flanked by the senior boys school on one side and the junior boys on the other, with continual traffic between the two. The girls needed protection from boys at all costs! This archaic restriction was in keeping with the grimy old building covered by decades of soot, with a malodorous lavatory block in the playground exacerbating its depressing effect. Dark wood-block floors, and dreary green and brown paintwork, created an atmosphere of severity, perhaps appropriate for a serious institution, but also reflecting the character of the Headmaster; I never once saw him smile.

Thank goodness the two scoutmasters were very different; modest, dignified, and always cheerful. I lost no time in joining the school troop of seventy-two boys organised in nine patrols, which had been founded in 1915, only one year after Grandad's troop had been formed. Troop meetings began as the patrols came to attention and saluted, as scoutmasters and the troop leader entered with the Union Jack. I couldn't help comparing this dignified procedure to Grandad's meetings at Tetley Street, where boisterous noisy youngsters played about and never hesitated to pick up a trumpet and 'give it a toot'.

First, our Scoutmaster inspected personal hygiene, (including fingernails and ears), then uniforms, footwear, and neckerchiefs. The unique scout uniform conceived by Lord Baden-Powell ('BP') was based on his experiences in the Boer War, and I suppose the hat was designed to keep off the 'burning sun'. However, the damp English climate made it difficult for even the most fastidious boys to keep the large brim really flat, and if it happened to be sat upon the results could be disastrous. So, it was just as well that this inspection was held before any vigorous activities took place. Our neckerchief, about two feet square, useful as a First Aid bandage or sling, was collected under the chin by a 'woggle', rather like a napkin ring.

The woggle was the only part of the uniform free from regulation, and later on, at the International Jamboree in Holland, we found several specimens that had been beautifully made from carved wood. A neckerchief could be used for sending signals in Morse code, but resourceful boys found many other uses for it, such as polishing shoes prior to inspection—to the despair of mothers who had meticulously washed and ironed them to perfection.

Otherwise, the uniform was strictly regulated; patrol colours on the left shoulder, with 'status' badges, such as First Class and King's Scout badges underneath. Proficiency badges were on the right arm, with Ambulance badges at the top of each arm so that the red cross would be visible in an emergency.

These details were carefully watched, so inspections could take some time, especially as young boys couldn't be bothered to put leather polish on hat bands, shoes, and belts that were loaded with knives, wallets, whistles, and, in my case, a small hand axe. Being made of cheap leather, belts soon showed signs of wear when not regularly oiled and polished. To be penalised was considered shameful, because it meant a loss of points in the Patrol Competition, in which the winning patrol enjoyed privileges, such as choosing the best site for their tent at the annual troop Camp.

After notices had been given and any badges awarded, each patrol retired to its 'den' in one of the classrooms under the charge of its Patrol Leader. These 'PLs' were the backbone of the troop; usually sixteen to eighteen years old, in the sixth form and about to take their Higher

Chapter 2: 1931–1938 High School and Scouting

School Certificate exams, some with a view to going to university (a rare thing in those days). Most were Prefects, having responsibilities awesome to young recruits like me. Some heroic figures who played rugby or soccer for the school were regarded as Very Important People.

So when, at my very first meeting, one of these VIPs invited me into the Seagull's den, sat down beside me and started to teach me the Scout Law and Promise, which he said I must learn off by heart, I was thrilled to think I might be a member of his patrol. He told me that the uniform was exclusive to the troop, sold by only one shop, and could only be acquired by presentation of a note from the Scoutmaster—it was like being admitted to a secret society!

Many of the boys in the lower forms never spoke to a sixth former throughout their whole school career, unless they were hauled up before some prefect. Yet here I was, befriended by a senior boy, and about to meet the rest of his patrol who were all older than me. This camaraderie was extended when, in due course, I came to know and make friends with most of the seventy-two boys in the troop. Scouting made my schooldays a rich and rewarding experience, whereas, for most boys, school was mere drudgery from which to escape into the real world as soon as possible. I now realise that I was very privileged indeed.

Actually, this first meeting proved to be a 'honeymoon period', followed by a rude awakening, for there was a tradition of testing the metal of new recruits by teasing them. I was easy meat, for having had no siblings to deal with at home I was easily hurt, but I sensed my Grandad watching over my shoulder, and determined to 'stick it out'. It wasn't all that bad. At camp the youngest recruit slept by the door of the bell tent, with the job of opening and shutting it. The first time I had to do this, my PL, Eric Benson, waited until I was tucked up in my blankets (sleeping bags were unknown) and then told me to get out and slacken the guy ropes. This was necessary because if they were left taut, the morning dew would shrink them enough to pull the pegs out of the ground, and the tent, with its heavy pole, would collapse on to the sleeping boys. When I got back and snuggled down again my toes encountered a slithering grass snake. It was a horrible shock but didn't actually hurt me; it might have done some good by giving me, as Father would say, 'some backbone, my boy!'

It was the patrol leader's job to teach woodcraft, so Saturday patrol outings were spent in woods where fires could be lit and food cooked. I was taught axemanship, to find out which timber burned well and which was useless, and how to find edible mushrooms. I learned how to light a fire in pouring rain using wet wood and only a tiny amount of kindling extracted from difficult-to-find places sheltered from the rain. First aid was learned the hard way—on blisters, cuts, and bruises arising from scampering among rocks during pitched battles, using horse chestnut conkers, and acorns for ammunition. On our hikes to Druid's Altar on the heights above Bingley, or to old ruined buildings on the wilderness of Baildon Moor, we vied with each other to name birds and searched for wild flowers. A minimum of thirty species of plant was required for the Naturalist Badge and that number was difficult to find in the sooty environment of West Yorkshire.

In those years of economic hardship, miners' strikes, and short-term working, the troop had to get by on a shoestring budget. The school didn't charge any rent, so our subscription of around one penny a week covered most running costs and even provided subsidies for scouts who couldn't otherwise afford to go to the annual camps. The Scouters had to exercise extreme thrift, and the 'balance in hand' in the annual accounts varied from the equivalent of one pound in 1928 to twenty-three pounds in 1935. (I've recently learned, to my horror that in 2013, some English school troops were being charged rents of hundreds of pounds a year!)

When one of our ancient First World War army bell tents deteriorated beyond repair and had to be replaced, an army marquee was offered to us at a bargain price (we could eat our meals out of the rain—what bliss!) A sale of work was organised, and parents were beguiled into bringing friends and relatives to buy our homemade goods. This sale was taken very seriously; some beautiful leather goods were turned out with painted pottery, wooden utensils, cakes and cookies, and, of course, refreshments. One year, my patrol, the Curlews, made model ships, which sold surprisingly well. We not only raised funds, but also had fun and learned from our experience (American scouts raise money by selling boxes of commercially made cookies, but don't have the fun of actually making them). The marquee was a resounding success, transforming camp life during wet weather, of which we had plenty.

Chapter 2: 1931–1938 High School and Scouting

The 'Woodcraft Circle' appeared to be mysterious, but as I learned more about it I began to realise that it was inspirational and forged a bond of loyalty and affection lasting long after the boys had left the troop. It was based on the tradition in some North American tribes of giving a new name to any brave who had distinguished himself. So, becoming a member of the Circle was regarded as a great honour, to be awarded only after many years of service. Our scoutmasters were always referred to by their Woodcraft names. This served to lessen the awe with which small boys looked up to them. You couldn't be frightened of a man who asked you to call him 'Red Dove' instead of 'Mr Clegg', even though you knew he was the sixth-form master, and Deputy Headmaster.

The Group Scoutmaster, Ernest Knowles (a teacher of French who was famed for aiming pieces of chalk with unerring accuracy at any boy in his class who wasn't paying attention), always expected to be called 'Singing Hawk'. He had been one of the very first scouts in the troop when it was formed in 1915, and then an Assistant Scoutmaster before enlisting in the Royal Engineers and being wounded while serving in France. He always supervised the camp kitchen. We had large, corrugated, galvanized bins for steaming puddings such as 'spotted dog'—a mixture of dough and raisins, which, if not well cooked, (as happened all too often) was dreadfully indigestible. They were the sort that Mother used at home for her domestic washing, and for steaming salmon they were perfect, as long as there was plenty of water in them, but if they were left over a cooking fire and allowed to boil dry, the solder would melt and the bottom would fall out. Ex-army cast iron dixies withstood any ill treatment, but they were extremely heavy and it took two youngsters to lift one when it was full of stew or porridge.

From 1933 to the outbreak of war in 1939 I never missed a camp—a week at Whitsun and a fortnight in the summer. Cooking duty came around one day a week, with eight boys catering for fifty or sixty people. This was pleasant enough in dry weather, but I well remember having to get up early to cook a huge dixie of porridge over a smoking fire, with rain coming down, eyes streaming from the smoke and the porridge so stiff it was hard to move the spoon. This job was invariably allotted to the youngest tenderfoot who rarely had the strength or tenacity to persist

'Singing Hawk' supervised the camp kitchen.

in his efforts for as long as an hour or more. So the porridge was usually burnt until Singing Hawk introduced 'haybox cooking' in which the porridge dixie was heated the night before and then put in a tea chest with hay packed around it, to cook until breakfast. This system was so much easier, and the results were delicious. It's a pity he didn't discover the method during *my* time as a tenderfoot.

A further hazard was the interest taken by wasps in the sixty or more slices of bread and jam, usually handled in the dark recesses of the cookhouse. An isolated and depleted jar of jam enticed the wasps to stay near the tent door, but the inevitable finally happened, and in the dim light of a rainy morning one of the recruits spread a mixture of jam and wasps over a whole tray of freshly buttered bread.

Apart from working parties to collect firewood and clean up the site, cooking was the only real chore, and the boys were free to enjoy any activities they might devise. Highly organised games involved everybody in camp, with stalking, hiding, and chasing, under rules that required umpires to be appointed. Camping was such a marvellous experience, that one year we invited four schoolboys whose parents couldn't afford to give them a holiday. It was a disastrous idea; they had no sense of discipline. One boy in particular just played the fool and tried to making people laugh by throwing up his hands and pretending to drown. I was on lifeguard duty and realised that he was taking a bit too long to resurface, so I dived in and fished him out of the deep part of the river into which the current had taken him, well out of his depth. Believe it or not, the very next day, while pretending to be a monkey, he fell out of a tree and broke his ankle! We were relieved to have an excuse to send him home. The other guests were not such a liability, but none behaved well enough to make us repeat the experiment.

At another camp, at Hornby, Lancashire, I took a few boys for a walk down the riverside, to see who could throw my javelin the furthest. The river Lune is famous for its salmon, and there, in a quiet pool, was a beauty, basking in the shallows. I flung the javelin in its general direction, and to my astonishment, I hit it. As we dragged it to the bank it dawned on me that I was supposed to set the younger boys a good example and here I was—poaching salmon! We hid it under an overhanging branch, and on the way back to camp laid a trail of clues. Singing Hawk announced that some poachers had been surprised downstream, and the boys were to 'follow the trail and search for evidence of any crime'. It was a struggle to get the salmon out of the water—it weighed sixteen pounds and was very slippery. We steamed it for supper, and the farmer happily accepted a large portion for his family, promising not to tell anybody.

Another unforgettable incident arose in the tiny church at Silverdale, where the miniscule choir of six was dominated by a very tall lady wearing a hat with a net veil. As she sang, the veil was first sucked into her mouth, and then blown forth horizontally in such a way as to convulse the audience of small boys desperately trying to stifle their giggling under the stern glare of the embarrassed Scoutmasters. This entertainment was repeated every time the choir sang, even the anticipation of this being enough to start the boys off again.

At summer camp the seniors always built a huge campfire, to be lit ceremoniously at sundown by the youngest tenderfoot, and followed by an entertainment of playlets, monologues, and music, orchestrated by the campfire leader. When the embers had died and the last chorus faded away, a group of shrouded figures would approach silently from the shadows. The leader would announce that this was a meeting of the Woodcraft Circle and would relate the story of the war between the

Sioux and the Iroquois, in which a young brave saved the Sioux from defeat and was honoured by being given a new name: 'Swift Arrow'. He would then describe, in flattering terms, the attributes of someone in the audience, taking great pains to avoid clues about whom he was speaking, so that most boys were kept guessing who was going to be invested with an Indian name until the last minute.

Some names were inspired. 'Red Dove' described Russell Clegg perfectly; a rotund figure with a florid complexion, and the mildest nature. 'Singing Hawk' epitomised Ernest Knowles, with his caustic wit, impatience with incompetence, but a generous nature and beautiful singing voice. These two Scouters were the backbone of the troop and a perfect combination of executive skill on the one hand, and resourcefulness on the other. Ken Emmott and I were invested at the same ceremony—he was named 'Black Arrow' for his prominence in athletics and I became 'Black Badger'. Another hallowed tradition was the 'tent feed', with an extension to the normal lights out, and a promise to keep voices down. The usual menu of cold baked beans, potato crisps, chocolate biscuits and pop, was consumed while lying luxuriously in bed, followed by tinned fruit embellished with condensed milk. It proved to be a test of the digestive system, which some of the younger scouts failed. Tent feeds invariably closed with the telling of a ghost story by an Assistant Scout Master, whose secondary function was to ensure that the time limit was not exceeded. The only story that I can recall is 'The Iron Maiden'. A tourist in a Transylvanian castle accidentally dislodged a stone that fell near a cat playing with its kittens, killing one of them. His girlfriend was very distressed, and he took her down to the torture chamber to distract her by demonstrating how the 'Iron Maiden' worked. This was a coffin, with a lid suspended from above by a rope and pulleys, having long spikes protruding from it, designed to penetrate the victim's body as it was lowered. To make things more realistic, the stupid man laid himself in the coffin and persuaded his girlfriend, much against her will, to unfasten the rope and slowly start to lower the lid. At this point the cat entered the chamber and leapt up to scratch the girl's face, at which she let go of the rope … a story that could be elaborated with the most gruesome details. I often wondered how the youngest boys avoided having sleepless nights after hearing it, but they were usually sound asleep by the time I got up to leave.

Scouting was a wonderful experience for boys without siblings, like me, teaching them how to interact with other boys. Otherwise my time in school ran far from smoothly. The Headmaster promoted me from Form I directly into Form III. The purpose of this experiment was never explained, but it had a disastrous effect on my schooling. I missed Latin altogether, vital sections of French and Mathematics, and also basic lessons in metalwork and woodwork that would have saved hours of time learning by trial and error many years later. Moreover, I was regarded with resentment by the existing members of Form III who were older than me and looked upon me as a young swot. They made it clear that I would be accepted into their society only after proving myself.

How to prove oneself? It seemed that this could only be achieved by taking on dares such as stealing sweets from the tuck shop, or Dinky toys from Woolworths, or by kissing one of the girls from the girl's school in broad daylight. For days I pondered, and finally, in desperation I took a Bunsen burner from the Chemistry lab and popped it into my satchel. It was an awkward trophy—shaped like a candlestick, and of no earthly use to me for we had no gas supply at home. I wondered what the chemistry master would do when he found one of his burners missing. So, I sweated and trembled a little, but by the time I got home I was quite euphoric about my newfound daring and dashed upstairs to show my trophy off to Mother. She was horrified, moaning, 'my son is a THIEF!' and crying inconsolably. I tried to explain that this was not stealing, but just a prank, but nothing would move her until I promised to return it the very next day. I didn't get much sleep that night, wondering how I could get it back unobserved, because the laboratory doors were kept locked, and I had no business being anywhere near the lab until the following week. In the event I was lucky; furtive pressure on the doorknob revealed the door open, a quick peep inside and a dash to the nearest bench got rid of the damn thing, with my unwavering resolution that I would *never* get into such a mess again. I didn't tell anyone about it; I was too ashamed and embarrassed. From then on I didn't care 'a tinker's cuss' what the other boys thought of me.

(This was a valuable experience for me, because much later when I sat with the Patrol Leaders on a Court of Honour, debating misdemeanours of scouts, who might have broken the Scout Law (if not the law of the land as well), I reflected how easily little boys could be misled, and was able to persuade the Court to make allowances while meting out punishment. Verdicts of the Court were treated with great respect, and penalties were usually mild, because expulsion from the troop was regarded as a failure on our part. Later on, two boys were turned out of the troop because they were a bad influence on younger boys, and ended up in juvenile court).

So far as I can recall, I was naughty only once, after the boys at the farm near Marrick Priory asked quite seriously 'Have you seen the Ellerton ghost yet? It's moonlight tonight, and it only appears at midnight'. Ellerton Abbey was a ruined mediaeval nunnery about a mile down river from our camp in Swaledale. We had an anxious debate as to whether ghosts really existed, but it was a challenge we couldn't resist. That night three of us crept out of our tent and made our way downstream, while the camp slept, we hoped, for this was highly irregular and we gazed at the ruins from the shelter of a wood. Billy Wright had a luminous watch, and as midnight approached we became increasingly nervous and kept motionless and as quiet as we could. In fact the silence was awesome and I was beginning to think of retreating when there was a sudden ear-splitting screech that froze our blood and made my hair stand on end. Only then did we see an owl silently gliding away with a rodent it had grabbed from just under our feet. With a chill still down our backs we abandoned the nuns and hurried back to bed, our adventurous instincts more than satisfied for a long time.

I was thirteen when a popular PL, Jack Law, invited me to become Second of his Patrol. His parents were members of the First Church of Christ, Scientist in Bradford (our family attended the Second Church). I switched loyalties without any compunction, but it seemed that Jack never attended Sunday School, so I pleaded pressure of homework and that ended my formal religious education. Mother and Father were happy; they 'didn't think much of the people at First Church anyway'.

Jack and I became firm friends, and he took me, together with 'Fungus' Foster on a never-to-be-forgotten bicycling tour of the Lake

Chapter 2: 1931–1938 High School and Scouting

District in 1934. Fungus's real name was also Jack; he was so nicknamed because he was studying Botany. He was conspicuous because he attended the girls' school (nudge, nudge) for the botany classes, and as the regulations forbade any boy to speak to a girl within the school precinct, this arrangement had required his special dispensation. For this trip, I had to buy the largest pannier bags I could find to carry a tent, groundsheet, cooking pan, bedding, and food. Some gear had to be hung over the front wheel, dangling from the handlebars. The whole contraption was so cumbersome that it was difficult to get up enough speed to keep my balance on a level road and it was a real struggle to push it over Wrynose Pass! This was a steep climb of 1,281 feet with unremitting hairpin bends but with marvellous views from the top. It was an ambitious trip for a thirteen-year-old boy, but we got home again safely.

Actually, the only bicycle accident that ever happened to me was much later, on my daily journey to school. There was a sharp bend at the foot of a steep hill that I always took at full speed. One morning a vehicle overtook me halfway down the hill and forced me onto the loose gravel at the edge of the road. It was impossible for me to get back onto the tarmac without skidding and the brakes were useless on the gravel. I aimed for a hedge and sailed headfirst over it, landing on my back on the soft earth of a market garden. I was lucky to find my precious bike still roadworthy, in spite of a slightly buckled front wheel (helmets were invented fifty or sixty years later).

In due course, Jack Law was promoted to be troop Leader, and I was made PL of the Curlews. For a boy of fourteen, this was a wonderful experience, full of interest and challenges. I had passed the university entrance exam and was a prefect in the Sixth Form. Up to that time my King's Scout Badge and 'All round Cords' were generally regarded as a sufficient adornment, but my ambition was to win 'Gold Cords' (requiring eighteen proficiency badges) and the Bushman's Thong, which called for five more. In all the troop's existence, no-one had achieved this. But there were others aiming at the same target: John Hudson, a bespectacled boy, who I regarded as a 'bookworm', and Ken Emmott. Ken and I became close friends and we both swam for the school. He was in the first soccer team, the first XI cricket team and also played tennis for the school. I only achieved the first rugby XV, but we kept pretty level in our scouting careers. We went on our first-class journey together, cycling up to Bolton Abbey and mapping the route as we went, so that subsequent explorers could follow in our tyre treads; we were both invested as patrol leaders at the same ceremony in 1936. The following year, Ken, John, and I were lucky enough to be in a party of six from our troop selected to attend the International World Jamboree in Holland.

The journey to Holland started inauspiciously by an eight-hour voyage over the extremely choppy North Sea, on board a flat-bottomed paddle steamer so small that the captain had to keep moving the passengers about to keep the paddle wheels (one on each side of the boat) in touch with the water. We arrived at the Hook of Holland somewhat nauseous, but we were restored by a reception that publicity agents dream about. It was incredible; Jamboree fever had gripped the Dutch! The engine driver and porters welcomed us, and as the train trundled towards Vogelenzang, everyone in the fields looked up and waved to cheer our passage. The next morning, we found ourselves in a forest of flagstaffs, and a medley of tents, towers, entrance gateways, and gadgets of all descriptions. My report for the troop magazine complained that my 'foreign languages' were not much use; French scouts were far more interested in practising their English than in trying to decipher our halting French. My clumsy German was useless as Hitler had banned Scouting from all German-speaking territories. We acquired Dutch clogs, Jamboree pennants, Hungarian feathers, and foreign stamps. Swapping badges and mementoes, such as woggles by means of sign language, proved to be great fun, but we had nothing worthy to offer to Icelanders in return for their beautiful polar furs. The Americans camped in luxury with camp beds, mattresses and elaborate kitchen utensils. Dutch camps were models of neatness and efficiency—French camps the opposite. Campfires with Polish folk dances, Chinese sword displays and dragon dances, playlets in mime, and Scandinavian folk songs were enthralling. The Chinese and Japanese scouts camped next to one another and were on the best of terms despite the fact that their countries were locked in a very nasty war at the time.

Our adventures started when we began to explore the hinterland on hired bicycles, which had back-pedal brakes (quite new to us). The

Chapter 2: 1931–1938 High School and Scouting

manipulation of these in congested right-hand traffic required intense concentration, and when we saw 'Eager Chamois' disappear between a tramcar and a canal that was covered with green slime, we held our breath until we found that he had crash-landed on the very edge of the water. The Dutch police were impressive, with tassels and braid, high boots and revolvers, and they all spoke English, French, and German fluently. When we innocently steered across the middle of a road junction in Haarlem, a policeman careered after us on his bicycle, blowing his whistle and shouting—only to explain, in the nicest possible way, why we shouldn't have done so.

On our final day, the citizens of Utrecht took 2,000 scouts into their homes to entertain them, and, in Amsterdam, a stranger took us in and gave us cakes and cider; I shall never forget all this hospitality, nor the old lady in Bilthoven, her cottage covered in Union Jacks, who waved excitedly to us as we marched by.

It was while we were at the Jamboree that we learned of my appointment as 'Duke of Kent's Messenger'. The Headquarters of the Boy Scouts Association needed more money, particularly to help handicapped scouts, and the Duke had agreed to launch a national appeal. This was something quite new, for tradition held that Scouts should never beg but should earn their keep, and as Britain had been in a severe depression for many years, such an appeal was especially tricky. There was no television of course, and the idea was to get as much press publicity as possible by investing messengers to convey the Duke's Appeal from London to all the regions in Britain.

I was given an enthusiastic send-off to London and, for the first time in my life had breakfast on a train, the latest high-speed 'Yorkshireman'. This had one of the new streamlined 'Pacific class' locomotives, sister to the 'Mallard', which even today holds the world record for steam trains, with a speed of 126 mph. On the London Tube, I followed the County Commissioner's strict instructions to go by an Eastbound train on the Inner Circle to the Mansion House.

Here, in the reception chamber, with crystal chandeliers and walls and ceilings embossed in gold, I joined fifty-six King's Scouts, representing counties from all over England, Scotland, and Wales. The hum of conversation in different accents reminded me of the jamboree.

The Lord Mayor, Sir Henry Twyford, explained the desperate need of funds, and Lord Somers, the Deputy Chief Scout, dealt with the special needs of handicapped scouts.

HRH the Duke of Kent pointed out that it was appropriate to inaugurate this appeal now, to ensure the future of Scouting while Robert Baden-Powell, who inspired the movement, was still with us (he died in 1941). He told us 'When your county has raised £500, Headquarters should be notified and a blue pennant bearing the name of your county will be hoisted on the 'Discovery' (his personal yacht). When you have reached the £1,000 mark, a green pennant will replace your blue pennant. At the conclusion of the appeal, your county will keep your pennants to commemorate the appeal'. Such high hopes!

With due ceremony, all fifty-six messengers received their red pennants from the hand of the Duke and hurried to the tube station, now in the middle of the 'rush hour'. My directions had been quite clear—obviously I must catch a train going west to get back to King's Cross. The platform was overcrowded and the carriages were full to capacity, but the crowd relentlessly pressed through the doors until the standing passengers, including me, were packed like sardines. I had never experienced anything like it! We seemed to be taking an inordinately long time to reach King's Cross, and I managed to manoeuvre to a point where I could see a diagram of the Inner Circle Line. To my horror I found I was going in the wrong direction; I should have been on an east-bound train! The circle was an oval with the Kings Cross and Mansion House stations quite close to each other so on my first journey, the east-bound train had gone around the bend and was actually travelling west when I got off at Mansion House. By now, we were nearly halfway round the circuit, and twenty-two miles away from King's Cross, but nobody could tell me whether it would be quicker to change onto a train going the other way, or to stay put. So I stayed where I was and sweated in an agony of suspense, wondering what on earth I would do if I missed the train to Bradford, the last of the day. How could I get in touch with my parents? They were not on the phone, and I would have to spend the night at King's Cross station. In the event, I caught the train with a few minutes to spare. It was just as well that I did, because when I stepped down on to the platform in Bradford, I was astounded to find a civic reception waiting for me! With the Lord Mayor were the Commissioner, and civic dignitaries together with Red Dove, Singing Hawk, reporters, and photographers. With no idea of what to do, I saluted and handed the pennant to the Lord Mayor, who made a speech, and then handed it to the Commissioner who seemed to think the whole thing had gone very smoothly.

At morning prayers the next day, the Headmaster asked me to tell the school about the trip. He had spoken to me only once before, when

The author as Duke of Kent's messenger.

he called me up to the dais to accuse me, in front of the whole school, of being a raving lunatic. I had declined to attend some special evening classes that he was holding to coach students in French for the matriculation exam. This was ironic, for he was the stupid man who had made me skip Form II, when I would have comfortably learned the French verbs that he was now so eager to teach during one or two evening sessions. Anyhow, I managed to pass the exam by sticking to the present tense, which doesn't say very much about the quality of that exam paper.

'Lex', the County Commissioner, was a prominent Leeds lawyer who certainly did his best to publicise the appeal, taking me to his Luncheon Club to meet his business friends, and arranging for me to parade the pennant at all the public scouting events in the county. But the Scouting organisation had no experience of begging in this way, and parades and boring speeches made little impression on a community in the depths of a recession. So, fundraising didn't come up to expectations, and, much to my relief, the idea of returning our pennants to the Duke on board the 'Discovery', moored on the Thames near London Bridge, was abandoned.

The year of 1937 was notable, but in 1938 there was the prospect of a trip to Switzerland for scouts whose parents could raise half the cost, which was ten pounds (about twenty dollars) per person. That was a lot in those days, and it was decided that if parents could only raise half of their share, their son would be quietly subsidised, if only the troop could raise sufficient money. We followed our traditional way of raising funds, but as this Sale of Work required an exceptional effort, the Deputy Chief Scout, Sir Percy Everett, came up from London to support our publicity. My patrol decided to sell lemon curd. The recipe came from Mother who made it regularly, and we soon had the rich (and expensive) mixture of eggs, butter, sugar and lemon juice in a large cast-iron dixie heating over a primus stove, being stirred laboriously with a huge wooden spoon. It was ready to be poured into the heated glass jars when Red Dove asked us to put everything on hold so that Sir Percy could witness this critical process. So we waited, and waited, and the precious mixture got thicker and thicker and ever more difficult to stir. At last, in response to our desperate entreaties, he arrived and watched us heroically struggling to pour the glutinous stuff.

Chapter 2: 1931–1938 High School and Scouting

The next day was Sunday, and we had visitors to tea. When Mother, announcing my culinary achievement with some pride, dipped one of her best knives into the lemon curd she found she couldn't get it out again, and eventually the blade snapped in two! I don't know what other customers experienced, but the rest of the Sale of Work was a success, and the trip was on.

As it was to be a 'once in a lifetime' trip, I decided to be wildly extravagant. Colour photos were in their infancy. 'Dufaycolor' films for cameras, first marketed in 1935, were now on sale in Bradford. The snag was the price. Three shillings and four pence for six exposures made them very expensive, and a further one shilling and sixpence was required later to develop them. A very costly decision for a schoolboy with only one shilling a week pocket money. These films had a speed of only 10 ASA—one third the speed of a normal black and white film—and as Father's Brownie box camera had no controls over shutter speed or aperture, the conditions would have to be perfect for every shot. I raided my savings and bought two rolls. I explained to those being photographed that they would have to hold perfectly still for the shot. Luckily everyone took my efforts very seriously, and ten photos survived.

For ten dreary hours we trundled along, staring at monotonous flat French fields that seemed to have no boundaries, but our fatigue was forgotten with the sudden glimpse of mountaintops way above the clouds. Switzerland at last!

Our base was at Wilderswil, near Interlaken, and when the clouds cleared, we saw the Jungfrau at the head of the valley. Its icy peaks reflected the last rays of the sun, framed by a valley in darkness and the deep blue of the sky. The next day we climbed above the snowline to a hut, where we were allowed to borrow skis. We had fun on the beginner slopes, with valuable lessons on how to stop before getting into real trouble. After that excitement, the walk home down the cart track, that meandered gently down to the valley, proved too boring for one group of us. Hungry for their dinner they decided to take a short cut through the woods down the precipitous slope. A mistake; they arrived long after we had finished our supper!

The scenery never failed in magnificence, and field glasses were in constant demand, especially if Swiss girls were about. In an age when all magazine pictures were black and white, we had never imagined such beautiful golden-brown complexions.

We were also impressed by the kindness of the Swiss people. One lunchtime we were waiting for the sweet course, and as the minutes dragged by, Red Dove got more anxious, and finally told the manageress that we should have to skip the dessert, or we should miss the train that was to take us to Interlaken. She threw up her hands and protested that this was a special treat for the boys, made of meringue, ice cream, fruit, and cream, and it had to be eaten fresh. But we were not to worry; she would ring the station and get them to hold the train for us! The superb concoction finally appeared and it was heart-breaking to have to wolf it down—but we needn't have worried, for as we scampered into the station there was the train, with people hanging out of the windows to see what the delay was about (Swiss trains famously run on time), and they actually smiled while the courteous conductor ushered us aboard. As he climbed up, Red Dove turned to ask him if we should be in time for the boat at Interlaken. 'Oh!' he said, 'Don't worry—I'll ring the boat and tell them to wait for you'. Of course, Red Dove did worry, having no idea how to get from the station to the jetty, but there on the platform was a sailor sent to escort us to the boat, which was being held back for our arrival!

The following days were a whirl of activity, on railways up the sides of mountains and through tunnels cut deep into the mountainside. The Grindelwald Glacier, nestling almost on the slopes of the Jungfrau itself, seemed small until we reached it, by way of a tunnel cut under the ice itself. We sat on the ice in the glare of the sun and listened to Red Dove talk about 'Terminal Moraines'. After all, he was the Sixth Form geography master!

After the assistant scoutmasters had been taken up the Schynige Plateau by a friendly Swiss Scout from the village, the patrol leaders got permission to go there by themselves. We followed the railway track that occasionally disappeared into tunnels under the snow and we found ourselves walking past the tops of railway pylons enveloped by drifts that were twenty or more feet deep.

We visited the Aar Gorge and its surging torrent, and the Reichenbach Falls, crashing with a thundering roar that drowned all conver-

Chapter 2: 1931–1938 High School and Scouting

The Belle Vue scouts in Switzerland.

sation. At Murren, we gazed with envy upon a lucky group of people learning to ski, and spent an hour of ecstasy sliding down smooth, steep snow slopes on our lightweight waterproof capes. We went quite fast and were amazed at being able to keep dry, for the snow was hard and crisp and did not melt at all; it just would not pack to make snowballs. Not that we came home without having a snowball fight; the day of departure dawned on a blanket of fresh, new damp snow and a pitched battle marked the end of our wonderful trip.

By the summer of 1938, I was qualified to go to a university, which had to be the nearest one in Leeds. My hopes of studying to be an architect were dashed when it transpired that the Parkinson Scholarship (which was to cover my fees at thirty pounds a year for three years) could not be used for the School of Architecture because that was at the Technical College. The next best thing would be to take Civil Engineering. By living at home, the only expenses would be in the way of books, drawing equipment, bus fares, and lunch money. These items consumed all my available cash, and at the time I graduated, I owed Father some eighty pounds, for which he generously forgave me.

The Head of the department was a Senior Lecturer, Dr R H Evans, a dynamic and ambitious Welshman working for a DSc that required him to publish several scientific papers. He needed students to help in the research work, but the intake each year was only about half a dozen students, which limited his options. Of course, I hadn't the slightest idea of this at the time, and I was very flattered when he suggested that my exam results 'were good enough for me to skip the first year and go straight into the second-year course'. He explained that this would leave my third year open to do research 'which might possibly form part of a PhD degree'. Good grief! It was a dazzling prospect for a green freshman, and I promptly forgot all about the stress that my old Headmaster's decision 'to skip the second school year' had landed me in! Would I never learn?

It transpired that there was some logic to Evans' proposal, because the first-year introductory programmes were easy, being designed for school leavers who needed to be 'eased into' higher education. My troubles arose when I was faced with calculations involving internal combustion engines, electric motors and dynamos, for, having missed the introductory lectures, I had no idea how these things worked. Somehow I muddled through and also managed to squeeze in the subsidiary subjects of Economics and Geology. I actually enjoyed surveying and structures, for Evans was a gifted teacher who stimulated his students, and I soon developed an affection for him because of his kindness and generosity to students, who were 'not all that bright'.

Chapter 2: 1931–1938 High School and Scouting

The daily bus journeys to and from Leeds were made pleasant by fellow travellers. Ken Emmott was taking Mechanical Engineering, John Hudson a degree in the Arts, and Sam Garton, Civil Engineering, and we all caught the West Yorkshire bus at the end of Idle Road. Sam was a very modest, shy boy who became notorious (and very popular) after being brought up before the whole school at Morning Prayers and severely chastised by the Headmaster. He announced that Sam had been seen outside the school main entrance, in broad daylight, actually speaking to a member of the Girls School! The Head made it clear that such disgraceful behaviour was not to be tolerated. Sam was now a university student and happily free to consort with his girlfriend without any restraint.

With my overcrowded academic schedule, I hadn't much time to spare for social activities, but felt obliged to attend meetings of the Scout and Guide Club until they proved to be a waste of time. I wasn't interested in any of the girl guides, although this seemed to be the only opportunity to meet the opposite sex. Victorian conventions still pervaded student life. The newly opened Student Union Building had separate common rooms and the Old Refectory had separate dining rooms for men and women. In the large common room under the Great Hall, students could mix freely, but even there I found that Engineering students tended to keep to themselves during morning coffee breaks. Union Balls, dances, and socials were useless to me—I couldn't dance and had no time to learn.

For exercise, I joined the Boat Club that had a boathouse at Swillington Bridge near Leeds, where the River Aire was polluted with lumps of black sewage sludge floating on the surface that stained our 'whites' remorselessly. This state of affairs was inexcusable, for sewage disposal systems could be very efficient; the Bradford sewage plant at Esholt was world famous for cleaning filthy sewage to a point in which the final effluent was crystal clear and could be drunk safely. Not only that, the wily Bradfordians proved that 'where there's muck there's money', making a profit from the fats and grease they extracted from the sewage, derived from the washing of sheep's wool. Their products included cosmetics and face cream. In spite of soiled linen, after three very strenuous races on the river below the famous castle, we won the

Maiden trophy at Durham; I was left in a state of euphoric physical exhaustion with a pewter pot and a firm resolve to explore a less-stressful form of exercise in future.

At the summer camp in Nidderdale, we had fun building a bridge over a dry stream bed using small tree trunks to teach the boys how to lash timbers together. The rather portly Red Dove provided an enter-

Enjoying sculling at Swillington.

taining performance testing it and declaring it 'safe'. Of much more use was an elaborate pulley-way, devised to lift water up a precipitous bank between the river Nidd and the road, which saved a lot of spillage and bad temper, because carrying a full bucket up a steep and slippery slope was fiendishly difficult. I began to think that I might enjoy being a civil engineer after all.

Under the threat of hostilities, I spent the summer vacation working with Sam Garton in the Borough Engineer's office in Brighouse, scheduling houses that required bomb shelters. They were free to those who earned less than five pounds a week. Being sunk in a garden, Anderson shelters were damp and liable to flooding, so we were very relieved when Morrison shelters were introduced in 1941. These were steel boxes that could double up as a table and were capable of withstanding the collapse of a two-storey house.

When it was fine enough, Sam and I surveyed a road due to be realigned after the war; we were told to make a good job of it. It was more rewarding than surveying for shelters, as was my very first pay packet of thirty-five shillings. I shall never forget the thrill of spending two pence that I had actually earned on a bar of Cadbury's raisin chocolate, which had never tasted so good! At the end of the vacation, I started my 'final year' coursework just as war was declared and when blackouts and air raid warnings made normal life impossible.

The troop was then split into four sections, or 'dens'. I was still troop leader, but John Hudson was the assistant scoutmaster in charge of our den, and Father provided wonderful accommodation for us in the basement of his business premises in Bolton Lane. We had three patrols and our first contribution towards the 'war effort' was to collect salvage—mainly paper, but also aluminium and any metals that could contribute towards making aircraft or tanks. We then embarked on leatherwork that was sold at a whist drive, with proceeds going towards a Services Comforts Fund. When Father took over an upper floor with an open fireplace, the den became very popular, especially during bad weather. At one meeting, we found a flood and shut down a tap on a burst water pipe, saving hundreds of pounds worth of his precious stock of powder he was acquiring, with difficulty, from South America.

Our first really useful public service occurred during a fog, which in those days could be formidable; thousands of domestic coal fires and factory chimneys were belching out sooty smoke, which combined with the mist to create 'pea soupers' that reduced visibility to ten or fifteen yards. Driving was nightmarish; windscreens became coated with greasy soot, which the feeble windscreen wipers of that time couldn't clear. Father's Rover had a windscreen that could be ratcheted forwards and upwards so you could peer underneath it—although with the fog blowing directly into your face it was punishingly cold to do so. Its single (hooded) fog lamp was aimed at the kerb about ten yards ahead, but when that kerb disappeared, one had to proceed very slowly, hoping to pick up another, preferably on the same side of the road.

One really foggy night, our Friday meeting was interrupted by an extraordinary commotion outside. We found that a huge double-decker bus had taken a wrong turning and was now stuck in the middle of the narrow, winding Bolton Lane. There were no kerbs at all, merely a grass verge and a ditch that might have toppled the bus onto its side and no lights anywhere because of the blackout. The hubbub was coming from a lot of gesticulating people arguing about what to do. It wasn't a simple problem; motorists recognised that bus drivers were reliable, so Bolton Lane was now jammed with a column of cars and lorries that had been following the bus! Not one of them was willing to try backing up the narrow lane, so after a lot of swearing, it was arranged that we should escort the bus to the bottom of the lane, with a boy walking on each verge, their torches guiding the bus round awkward corners that normally it would never have attempted to negotiate, even in daylight.

The next time we had bad weather, we decided to prevent this happening again by waiting in Bolton road and guiding buses across the junction where the kerbs were so far apart that the bus drivers could not possibly see them. These fog patrols were very much appreciated, and some years later I found myself guiding a big red London bus from the bottom of Whitehall, across Parliament Square to Victoria Street, followed by the inevitable convoy of taxis, lorries, and cars. On that occasion, the driver actually leaned out to thank me.

Devising outdoor activities for the boys was fun. I had learned at university just how simple 'plane table' surveying could be. There were no instruments involved; just a tripod with a flat board on top, together with paper, a ruler, and pencil, and it seemed a good idea for scouts to learn how to make a useful map. One Saturday we had just set up outside Eccleshill Church in Otley Road to practise, when along came a policeman. Apparently, 'surveying the country in time of war' was illegal; we might be passing information to the enemy. The policeman took down our names and addresses, and we thought that would be the end of it, until a journalist, Jack Blunt, wrote a hilarious article in the weekly magazine *'The Scout'* about the episode, with cartoons showing us being carted off to the nick.

I thought this to be ludicrous at the time, but, after the war, I was astonished to find that, further along that same Otley Road, was a huge underground factory immediately adjacent to the aerodrome at Yeadon where military aircraft had been assembled and flown to operational airfields. So we had inadvertently picked a very sensitive area for our surveying operations! The camouflage of this factory was impressive; its roof sloped down to ground level to avoid any shadows on aerial photos, and, during the war, trees were 'planted' and cattle distributed over its vast flat roof. The dam where Malcolm and I had sailed our boats had been camouflaged to make it invisible to German aircraft. Although hundreds of people must have worked there during the war, I never met anyone who knew that it existed.

Shortly afterwards Leeds had its first air raid. The following morning, on our usual journey to university, we found a house next to the Rodley bus stop completely flattened, and fires still smouldering along both sides of Kirkstall Road. In the centre of Leeds, the city museum had suffered a direct hit and burnt out, and on my way up to the university, we had to pass a flaming gas main in a crater just outside the Town Hall.

Life had begun to get really serious.

Chapter Three

1939–1945: The Second World War

I was now halfway through my degree and was told to forget about joining up and instead to concentrate on getting qualified, because the government wanted as many trained engineers as possible. All teaching courses were accelerated by eliminating vacations, so that the 'academic year' was condensed into nine consecutive months.

Dr Evans had enrolled me as a PhD student and manipulated my course so that I had jumped the first year; I was now making and testing pre-stressed concrete beams for my doctorate and swotting for final exams. I didn't know it at the time, but Evans had evidently persuaded the Selection Board that my research on concrete was essential to the war effort. He pressed me to produce a thesis setting out the results of my experimental work as soon as possible. Pre-stressed concrete would certainly have been ideal for the construction of the Mulberry harbour being made for the invasion of France, because it was stronger than ordinary reinforced concrete and uses less cement and steel, but this technology was still in its infancy.

By this time, France had been over-run by the Germans and I saw the first trainload of weary soldiers from Dunkirk arriving at Leeds City station. They were to be billeted in the newly completed Quarry Hill flats, at the bottom of the Headrow. Following the air raid on Leeds, I decided to live in the laboratory to make use of every minute of daylight, because blacking out the large laboratory windows was impossible. I joined the 'nightwatchers', who patrolled the university buildings during air raid alerts, ready to deal with incendiary bombs such as those that were devastating London, Coventry, and Birmingham. These duties included a free breakfast, which suited me very well. I worked in the main laboratory from first light and wrote up the results in a smaller room that had blackout shades. Lack of sleep became a serious problem, but I managed a first-class honours degree and had just started to analyse a mass of experimental figures for my higher degree when I was called up for War Service in London.

I joined the department Tank Supply 9 (TS9) with all the ardent fervour of a young man of nineteen determined to play a significant part in the war effort. The defeat of the French army and the retreat from the Dunkirk beaches had been a blow to our morale, redeemed only by the incredible performance of the RAF in the Battle of Britain. Like most people, I had the impression that our tanks had let us down. The Matilda (I wonder who thought up that name for a heavy battle tank) had been designed in 1934 and was intended to support troops in the sort of trench warfare experienced in WWI. It was heavy (twenty-five tons) and slow (15 mph) and the German tanks 'ran rings round them' for the Panzers were light (ten tons) and fast (25 mph), but their guns were no better than our two-pounders, and their armour was much thinner. The Matilda was tough, and even the Mark III Panzer's special armour piercing projectiles couldn't damage them, so it was only the unexpected tactics of the Germans that won the battle when massed formations of their panzers broke through the Allied lines and wreaked havoc with our supplies and communications.

The best Allied tank at that time was the French Somua S35 with a 47 mm gun, a speed of 25 mph, and even a radio—fitted as standard. Its castings of high quality made it expensive to make, but its main drawback was that the radio had to be operated by the commander at the same time as his other duties, which included loading and firing the gun. In contrast, the Panzer III had a three-man turret, where each man could concentrate on his own job. When the Panzers broke through Belgium in their hundreds, the 250 Somua tanks were spread evenly along the whole length of the front, with no time for them to be grouped together to meet the enemy.

Chapter 3: 1939–1945 The Second World War

TS9 had the job of extending factories to house machine tools for the manufacture of tank parts. These facilities were dispersed as widely as possible to minimise the effect of bombing—some were just back street garages. As I was the only technical member of the department, anything requiring drawing or designing came to me, and I was kept busy working an eleven-hour day, seven days a week, with alternate Sundays off. I was given special permission to work in the late evenings to write my thesis, which ultimately earned me an MSc degree (a PhD would have required three years' residence).

My duties changed over the years from designing pits for testing various methods of waterproofing tanks and even making them float, to the issuing of licences for materials, and later, to financing building work, so I only had a marginal knowledge of the vehicles themselves. The regimen of secrecy didn't help, but I remember the day when some high official decided that the term 'tank' (used in the first World War as a code word to hide our new invention) could safely be abandoned. His bold decision resulted in our being named FV9, so we were now exposed as the department responsible for Fighting Vehicles, and no longer disguised as plumber's merchants!

The need for secrecy was still paramount, of course, and, for months, I pondered the meaning of CDL, which appeared on some documents, but 'Canal Defence Lights' didn't mean any more to me than 'CDL'. This turned out to be one of the best-kept secrets of the war; over 200 tanks had been equipped with searchlights intended to blind or disorientate the enemy. As it happened, no battle occurred that suited their use, and it was just as well, because we heard that the Russians had used similar devices with disastrous effect to themselves. They made easy targets for enemy gunners who were not blinded by the searchlights. However, our CDLs did prove helpful to engineers building bridges, clearing obstacles, and recovering damaged vehicles in darkness.

By the end of 1940, the Germans had reduced the number of daylight raids on London, but bombs were still being dropped by night. The only lodging I could find in that badly battered city was an attic room in St John's Wood, lit by a skylight through which I could see the sky and little else, but that leaked rain and snow onto my bed. The house was overcrowded and tenants cooked their smelly meals on the staircase landings. So, I didn't really mind working long hours, and my Christmas Day 1940 was spent in the office. The Londoners' blackout in the winter of 1940–41 was not half-hearted, as it was in the North, and I needed a torch to grope my way late at night from St John's Wood Tube Station to my digs. One night, an unexpected air raid alert sent me scrambling for shelter, and I got hopelessly lost. At about ten o'clock no-one was about, and by the time I did find my lodgings it dawned on me that if I were ever caught in a London 'pea souper', finding my normal route would be a real problem. So, following the scout motto 'be prepared' I counted my steps as I left the underground station and evolved a formula along the following lines; R 127; R 92 to post-box; L over main road 25; R 42; L 20 to lodgings. I practised this with my eyes closed every night until I could walk without hesitation and had only to remember the number, but in accordance with 'sod's law' that winter remained free of fog. All I succeeded in doing was to make a lot of people wonder if they ought to offer some sort of help to this strange young man who was wandering about with his eyes closed.

My bicycle was invaluable during the war.

When I realised how valuable my bicycle would be, I asked Father to send it by train. Of course bikes, and their spare parts, were unobtainable, and I was in a constant worry that it might be stolen. Whether it was the extraordinary sense of loyalty and camaraderie that pervaded London during the bombing, or the fact that delinquents not in prison were in the Services, I don't know, but I found that I could leave my bike in the street with confidence, like many others who rode their bikes to work.

I found that Londoners would go out of their way to help anyone in trouble and were proud to

Chapter 3: 1939–1945 The Second World War

do so. With such poor housing, you can imagine my excitement when one of my colleagues at work recommended me for a vacancy in a hostel set up by servicemen in 1915 known as 'Toc H'[1]. All members were engaged in the War, and the communal spirit was quite wonderful. Over the front doorway was a prominent notice:

'ABANDON ALL RANK, YE WHO ENTER HERE'.

Everybody from senior officers down to lower ranks were known only by their Christian names, and never divulged their work, so the only clue to their rank was when they occasionally appeared in uniform. Membership constantly changed with postings, and, on average, only half the members were 'permanent' like me. Later, in 1943, when the pressure of work was reduced from seven to six days a week, everybody was expected to take on some social work, and, until 1946, I served as Secretary of the Westminster Club for the Blind, arranging speakers and organising their meetings.

However, in the meantime, the war dragged on interminably. I thanked my lucky stars to be in such a happy home, on the Thames embankment at the corner of St George's Square, Pimlico. The square had a central garden surrounded by large eighteenth-century houses like that occupied by Toc H. During air raids, some people had to cross the square to reach bomb shelters in the garden, but we had the 'luxury' of being able to shelter underneath the pavement in what were originally coal cellars; after any local bomb blast we were covered in coal dust. Rather than emerging in a state that needed a bath, most people, including me, chose to forgo the shelter and stay tucked up in bed, hoping for the best.

Department FV9 was housed in Adelphi House, just off the Strand, near the Savoy Hotel, overlooking the Thames. The morning number twenty-four bus from St George's Square was good, but late at night, when air raids were on, the Underground was a safer option. The only problem was to get from Victoria station to Toc H through falling shrapnel from a battery of anti-aircraft guns and rockets in Battersea Park, just across the river. Debris from their shell bursts clattered down on all the roofs and pavements and they could cause injuries. Fortunately, most

---

[1] Toc H was originally a soldiers' club in Belgium in 1915, named Talbot House after a young Lieutenant Talbot killed at Ypres. 'Toc' is the army signaler's code for the letter 'T' and 'H' is for house.

of the buildings in the district had substantial porticos in front of their main entrances, and it was just a question of earmarking one to which I would dive for cover next time a salvo of rockets was fired.

Our fire-watching post was next door, and groups of us took it in turns to keep watch from the cover of its entrance portico. One night, I was inside when we heard a stick of high-explosive bombs getting closer and closer; thump… thump… thump… THUMP… and the wall next to me shook so much that I thought we had been hit. It transpired that six large men who had been outside, firewatching from the portico, all dived through the front door at the same time—collapsing in a heap against my partition! You can imagine the subsequent comments. But we had all been very lucky because that last bomb had exploded in the street right outside our front door.

I shared a room with Pat Whelan, a charming Irish Guardsman. He had been a sniper in Belgium in 1940, and when Stuka dive-bombers attacked the village, he and his observer had run for the cover of a large tree. The observer was blown to pieces, but the tree trunk had sheltered Pat—apart from his leg, which was shattered. He limped badly, and fragments of bone and shrapnel kept working their way to the surface of his leg. He worked in Military Intelligence analysing reports about weapons that the Germans were suspected of developing and producing drawings to show what they might look like. Although we had both signed the Official Secrets Act, and were not supposed to divulge specific information, Pat's work was stressful and he felt he had to confide in someone. As we lay in our beds at night, I was the recipient of occasional misgivings and forebodings about Germans and their weaponry, and about the war in general, which, of course, I never mentioned to anyone.

Temporary bridges had been built across the Thames in case the main bridges were bombed, and, in February 1944, I was puzzled to see Bofors anti-aircraft guns being mounted on them, because such guns were only of use against low-flying aircraft, and the German bombers flew high over London to avoid the barrage balloons. When I raised this with Pat, he told me of the recent Mosquito raid on the prison in Amiens, where a low-level attack had released over 250 members of the Resistance, who were about to be shot by the

Chapter 3: 1939–1945 The Second World War

Barrage balloons helped protect London from low-level bombing.

Gestapo. The War Office suspected that the Germans were planning a low-level raid on Whitehall in retaliation. So, when the sirens went off a few nights later, followed by an almighty explosion, Pat was quite certain that 'this was it', for Whitehall was only about a mile away. The next morning, it was reported that an enemy bomber had crashed onto a railway line just on the other side of the river, but this report was misleading. It was indeed an aircraft, but it was unmanned, and turned out to be the first of the V1 flying bombs that became known as 'doodlebugs' because of the horrible rattling roar that came from their ramjet engines.

Each flying bomb contained one ton of explosives that could flatten a whole street of houses, and people found the roar of an approaching doodlebug terrifying. If they passed overhead, and went on their way, sighs of relief were audible, but if the engine cut out and it started to descend, making a swooshing sound, everybody dived for cover, for once the gyroscope ceased to function, the bomb was quite capable of descending in a broad spiral and catching people who thought it had passed them by.

As the enemy built more launching ramps, the numbers of flying bombs increased, and the sirens kept going through the night, making

A V-1 falling in silence near Drury Lane.

us dive under our beds with our sheets and blankets—for this was the winter of 1944 and our rooms were unheated. We stayed under the beds until we reckoned the danger was over, when we could pile the bedding on top the beds again and go to sleep. One morning I awoke as it was still dark, and as I lifted my head I banged it on a piece of iron only inches from my face. I felt blood trickling into my eyes and started to panic as I tried to roll from underneath the iron bar and then found that my legs were trapped. I was convinced that the house had been hit and that I was covered in debris.

I called out to Pat: 'Pat! Are you alright?'

A groan came from Pat, who was a heavy sleeper, 'Really, Geoff, it's not time to get up yet—it's still dark!'

Me: 'Have you got a light so we can see what's happened?'

Pat groped for his torch. 'Geoff, where on earth are you?'

'Under this girder!'

'Don't be silly—you're under your bed!'

Chapter 3: 1939–1945 The Second World War

It turned out that after the last V1 had passed over, Pat had got back into bed, whereas I had fallen asleep with my legs entangled in the bed sheets. It was embarrassing when I arrived at work with a bloody forehead—I mean what could I say? 'I hit my head on my bedstead.'

They would reply, with some disbelief, 'On your bedstead?'

As the weeks passed there was no point in sounding sirens because the V1s came regularly throughout the day and night, and the air-raid was continuous. Sirens were replaced by the roar of the doodlebug itself, which could be heard miles away or, in large buildings such as ours, by lookouts that sounded a local alarm if our building was directly threatened when we retreated to safe zones. Even these lookouts became superfluous when we discovered that the launching sites were made of concrete and immovable, and that the Germans adhered to rigid schedules for launching. Every day at 8.50 am, and again at 2.30 pm, we took shelter while the bombs, regularly aimed in our direction, thundered harmlessly overhead.

Although the number of flying bombs was increasing, Pat could not understand why we never heard them explode. It was not until after the war that we learned that Nazi spies, planted before the war in England and facing the choice of a firing squad or collaboration, were reporting to Berlin that the bombs were falling short. The Germans, therefore, put in more fuel, which they could ill afford, so that

London residents were forced to take shelter during the German bombing raids on London, and often the safest place was a Tube station.

the doodlebugs sailed over densely built up areas, and fell mainly on open countryside outside London.

At the same time, our latest fighters were finding it possible to catch up with doodlebugs although they were very fast. A Mark XIV Spitfire now had a speed of 450 mph (100mph faster than a 1939 vintage Mark I Spitfire) and the Hawker Tempest II was nearly as fast, at 435 mph. Extra fuel tanks allowed them to patrol for over four hours at a time and keep in touch with the ground radar that told them where the V1s were, so they could dive down and intercept them. But, as it was very dangerous to shoot at them, because an explosion in the air could easily destroy the attacking aircraft, the pilots evolved a safer method that required great skill. They flew alongside the doodlebug and flipped its wing so that the gyroscope lost control and the bomb hit the ground before it reached any built-up areas. Of the 1,770 doodlebugs destroyed by the RAF, 300 were credited to Spitfires, and 638 to Tempests.

London's defence was a concentration of anti-aircraft guns along the South Coast, backed by a balloon barrage all along the Downs. One Saturday, I resolved to visit this area on my bicycle in the hope of seeing a V1 brought down. It was a long way to go, but a beautiful day and I set off at the crack of dawn. I got home at dusk after a trip of eighty miles, having seen a lot of balloons but not a single flying bomb.

(Long after the war, I met Brigadier Ramsden when he became a Member of the Council of Leeds University. He told me how, on being promoted to Brigadier General, and put in charge of this very same balloon barrage, he made a tour of inspection to meet his officers and men. At one site, the young lieutenant in charge was asleep on his bunk. As his corporal shook him awake, he sprang out of bed exclaiming 'My God—the Brigadier!' Equally startled, Ramsden whipped round, saying 'Where?' They had a good laugh, and Ramsden, a delightful man, must have proved to be a very popular CO).

It wasn't until I got back into London that I saw some of the awesome carnage that had been caused by the V1s that had got through the balloon barrage on that Saturday. One street in Lambeth had been completely flattened, with the fire brigade and rescue workers still in action. When I reached Toc H, I found the area cordoned off by the police because a V1 had hit Dolphin Square, adjoining St George's

Square. I don't know how many people had been killed or injured—people never asked, and of course the figures were never published—casualties were classified as military secrets. But Toc H was in a real mess; windows had been blown out, doors off their hinges, soot and plaster everywhere. Mercifully, the house had been nearly empty, with most of my friends having been at work and the remainder, like me, enjoying the fine weather, so no one had been hurt.

During the spring of 1944, the invasion of France was well under way, and the launching sites for the V1s in the Pas de Calais were captured, thank goodness! However, V2 rockets, which were launched from the Dutch coast, were becoming more frequent; they were much bigger, did far more damage, and could not be intercepted. Their tremendous explosions were nerve-racking, but they were not as psychologically frightening as the V1 bombs, because, being supersonic, they arrived silently and if people heard the explosion they knew they were safe. The trouble was that it was not possible to stop them being sent, for they were launched from mobile vehicles. We had to suffer much more bombardment, and they killed 2,500 Londoners and injured thousands more before the Allies could over-run their bases in Peenemunde. It's curious that few people nowadays have even heard about V2 rockets.

V2 rockets, launched from the Dutch coast, were much bigger than V1s and could not be intercepted. By Carlos Moreno Rekondo [CC BY-SA 4.0 (https://creativecommons.org/licenses/by-sa/4.0)], from Wikimedia Commons

During 1941 and 1942 I worked a six-and-a-half-day week, with every second Sunday off. Until 1944, our leave was restricted to only five days a year. So you can imagine what a delight it was to be given an extra four days leave at Christmas 1944, when I went home and saw Mother and Father for the first time in over three years (I hadn't even spoken to them, for they couldn't get a telephone until the war was over). After a deliriously happy journey on a crowded train with standing room for most of the 200 miles, my wonderment at being safe in Bradford was indescribable.

I settled down in my own little bed and soon fell fast asleep. In the middle of the night I heard the drone of a distant flying bomb but took no notice for it was a long way away. But as I turned over my hand hit a wall—a wall? There wasn't a wall near my bed. I felt that wall very carefully and then recalled, of course, that I was at home in Bradford and not in London at all! So it must be a nightmare—but how was it that I could still hear that unmistakable noise? Was I in some kind of shock? Having delusions? Doodlebugs couldn't be anywhere near Bradford. I turned over and tried to go to sleep, but the noise was still there. I broke out into a cold sweat, certain now that I was mentally ill, and forced myself to get up and look out of the window. Thank God, there were at least four bugs scattering sparks over Bradford on their way, we learned later, to land on Manchester. It was the first time V1s had ever been seen in the North. The next day, the radio said that the bombs had been launched from bombers over the North Sea. Our fighter patrols made sure it did not happen again.

Back in London, and towards the end of 1944, war films began to appear, and I went to see one at the Victoria cinema. The men on the bridge of a destroyer were scanning the coastline and watching bombs coming down. One turned to the other and said, 'Jerry's busy tonight!' Almost immediately, there was an almighty explosion just outside the cinema; the lights went out and there was a long silence as the film had stopped. Then there was a roar of hysterical laughter as the audience saw the funny side of it, and we appreciated our lucky escape from that particular V2 rocket.

The air-launched V1s aimed at Manchester were the 'last gasp' from the Germans, and although still more V1s and V2s fell on London it was clear that the war would soon be over in Europe. When VE day finally came, I suppose I should have joined the crowds and made it a day I should never forget but my overwhelming sense of joy and relief was such that any time spent roistering in the streets would have spoilt it for me. I celebrated quietly by myself.

I shall never forget Toc H and all my friends there. Its premises had been 'adapted' rather than 'converted' and the rooms in what had been a town house were crammed to capacity. Pat and I had a room in the servant's quarters half way down the cellar steps but were privileged to have a tiny bathroom next to us. There was a gymnasium and a billiards table on which a massive Quarter-master Sergeant in a Guards regiment who must have weighed 300 pounds showed me how to hold the cue properly, and, oh, so delicately. In the basement was a chapel attended by everyone on Sunday, with its wonderful Ode of Remembrance, the Toc H prayer;

'They shall grow not old, as we that are left grow old,
Age shall not weary them, nor the years condemn,
At the going down of the sun, and in the morning,
We will remember them'.

FV9 was not involved in the war with Japan; the Americans had many tanks, so my last few months were spent at the dreary job of tidying up all the paperwork that steadily accumulated as people left on 'priority release'. As I had not been employed when I was called up, I was one of the last to be released, and, by then, jobs were becoming very scarce. I did not realise it at the time, but this was a blessing very well disguised. Promotions required a transfer to another department. Those people who were considered useless were promoted to get rid of them, and some incompetents rose to quite dizzy heights. People who were any good were kept firmly in their existing place.

Therefore, I am not ashamed that, five years after my initial appointment, I was released from the war service at exactly the same salary with which I had started, although my responsibilities had steadily increased and I was now doing work that the Deputy Director and senior executives should have been doing (had they not all managed to sneak away to better jobs). I had to wade through thick files to establish the financial state of dozens of contracts, and then present a complete picture to the Treasury. It was drudgery and I hated it, but I learned a great deal about how to deal with Treasury officials, whose only criticism of all my work was an objection to my coining the word 'Finalisation' as a heading for most of my reports. They said that such a word didn't exist, but finally admitted that they couldn't come up with a better one! I developed some skill in dictation, using an ancient wax-cylinder dictating machine, which prepared me to use the marvellous new machines that were then coming on to the market.

During the last weeks of the war, while Japan remained undefeated, I spent weekends in the countryside around London, using youth hostels. Then, having been given a whole week's leave to celebrate 'Victory in Europe', four friends invited me to cycle with them to Bath. I have never enjoyed such a hilarious holiday—everyone had his own special sense of humour, and we seemed to be laughing all day long. At one youth hostel, we assumed titles and watched with delight at the expressions on the faces of other hostellers as we referred to each other as 'M'lord', or 'Sir John', releasing the effervescent joy we all felt at the ending of that terrible war.

I certainly got my money's worth out of my old bicycle, and eventually sold it to a friend for five pounds; making a profit of sixpence over the fourteen years of ownership. I calculated that the frame itself—the only part of the bike that had not been replaced—had carried me between eight and nine thousand miles.

It was near Bath that a small plane screaming overhead, apparently on fire, startled us because smoke was trailing behind. It was not until many years later I realised that it must have been the experimental jet aircraft, the 'Gloster Whittle', powered by the Whittle turbojet engine, being developed in the hope that it would reach speeds capable of dealing with V1 flying bombs. A prototype had reached 466 mph in 1943, but the first operational jet aircraft, the 'Gloster Meteor', didn't fly until after the war.

VE day prompted Mother and Father to come south for a week's stay at High Wycombe, with Aunts Anne and Mabel, and also their old friend Mrs Swingler, who was in her late seventies. Nevertheless, she was the 'life and soul' of a very jolly holiday. Everyone was called upon to perform, and as I had never been to a 'party' I simply hadn't a clue as to what I would contribute. Mrs Swingler, bless her heart, saved my day by showing me a mime about a man painting white lines in the middle of a busy road. As others copied the mime it became increasingly funny and some said it was the 'event of the evening'. It certainly was for me!

Mrs Swingler offered palmistry and horoscopes and insisted on doing mine. She was very convincing, and foretold that I would become a lecturer. I was quite deflated, for my ambition to become an architect had evidently escaped her. But Mother was elated, for, to her, this could mean only one thing; that I was going to be one of those superior and gifted mortals who went about the world giving inspiration as a Christian Scientist lecturer. Mother returned home in high spirits.

It was while at High Wycombe that news came through about the dropping of atomic bombs on Japan. Unforgettable.

I persisted in my attempts to obtain employment, including an interview at the Atomic Energy Agency, which was about to build its first nuclear power station, but my lack of site experience was a handicap. I began to despair about my job prospects until, out of the blue, I got an invitation from Dr Evans to join him as the first lecturer in the newly formed Department of Civil Engineering at Leeds, to which he had been appointed as the first Professor. This was incredible! How on earth could Mrs Swingler have known? I jumped at the chance, without any idea of what I was letting myself in for, because, for the first exciting year, Evans and I would be on our own and running the whole course between the two of us. I was about to learn what hard work and high tension *really* were.

Chapter Four

1945–1978: University Life

Little did I think that Dr Evans and I would be founding a Department of Civil Engineering that would grow to have thirty members of staff plus seven professors, but the 'birth pains' we suffered in that first traumatic year might have given me some idea of the future.

For that first hectic session (1946–1947), the two of us had to manage all the teaching on our own. It was a formidable challenge, even for such a distinguished scholar as Rhydwyn Harding Evans, DSc, who was always known to me as 'Prof'. He was a dynamic Welshman, quick witted, impatient and with razor-sharp intelligence, who was building a formidable reputation for research in concrete technology. He was awarded Honorary Degrees by English and French universities and was eventually invested as a Commander of the British Empire. My problem was to keep up with him.

We were lucky to have Sidney Rider as senior laboratory technician. Bright and energetic, Sidney was Prof's 'right hand man' in everything, and quite indispensable, since there were only the three of us staffing the department and I was a lame duck struggling to keep up. Sidney was responsible for running the laboratories and making concrete beams to be tested. As a research student back in 1940, I had made beams myself, with meticulous supervision from Sidney, who carefully checked the proportions of cement, sand, and water as I laboured on the mixing floor with a huge shovel designed for use by a man twice my strength, until the 'slump' of the mixture was just right. It was he who showed me how to operate precision machinery in the Engineering workshop to make high tensile reinforcing bars for my experimental beams. My reunion with Sidney was a very happy one and I could understand why Prof said that it was a shame that he never had the chance to go to University, for he was as intelligent as the brightest of our students.

It was Sidney who, back in 1938, had given Prof two driving lessons after which he went to take the driving test! On his way there a large car emerging from the General Infirmary hit him. His old jalopy was rendered unroadworthy, and the other driver, a consultant at the hospital, full of apologies, insisted on lending his own car for the test. Evans had never driven such a large limousine with controls entirely different from his little car, but after listening to a hasty explanation of how everything worked, he rushed off for his appointment. Of course, being the man he was, he not only passed the test but was congratulated by the examiner on his performance! To celebrate, he gave Sidney his old car in return for the lessons. This was typical of his generosity; years later I learned how he was widely respected for helping students who were in financial difficulties.

The Prof was always in a hurry, and as a passenger, I found my foot repeatedly pressing an imaginary footbrake. One day, we were in his rather splendid Armstrong Siddeley, hurtling along an empty country lane near Whitby, when he slowed down because there were three or four sheep grazing at the side of the road. Suddenly Prof jammed on his brakes saying, 'That lamb is going to cross the road!' I hadn't even seen the lamb, grazing in the left-hand ditch, but sure enough, as soon as it saw the car, it scampered across to join its mother in the right-hand ditch. By swerving, Prof just managed to miss it. I never worried about his driving after that; I realised that I should never be such a safe and skilful a driver as he.

The Introductory Engineering lectures were attended by over one hundred students: civil, mechanical, electrical and chemical engineers, miners and metallurgists, including some women. You can imagine my embarrassment when, in my very first lecture, I turned towards the blackboard and found my trousers fly being ripped open by a gas tap that protruded from the demonstration bench. Zippered flies were unknown in those days and I had to stand there and casually button up again as though this was a commonplace occurrence. Not an auspicious start to my career.

Prof taught his favourite subject, Theory of Structures, and of course, supervised the Structures Laboratories. I helped in the Strength of Materials lab, the Hydraulics lab and the Drawing Office. I had no problems preparing lectures on Public Works, which covered railways, roads, water supplies, the design of dams, storm drainage, and sewage disposal, but along with the lectures on Surveying came a nasty shock. I found that I had to teach Geodetic Surveying, with measurements on stars, and calculations in spherical trigonometry about which I knew *nothing* at all! This was because, as an undergraduate, with only two years available to cover three years' work, I had found that the Surveying exam required answers to six out of ten questions, of which four were always about astronomy. So I gambled by concentrating on the other six, and by skipping Geodetic Surveying I saved a lot of valuable time. Now the chickens were coming home to roost, for while *learning* astronomy it is difficult enough, *teaching* it is quite another kettle of fish. I had always loved plane trigonometry but three-dimensional spherical triangles with curved sides, and angles adding up to more than 180 degrees were tricky. The familiar relationships became 'sinh', 'cosh' and 'tanh', in complicated formulae requiring the use of eight-figure logs instead of the usual four-figure tables. I now needed all the time I could find to learn about the subject, so I put off the lectures on astronomy until the late spring.

I was to have another rude awakening in the middle of March, when the printer demanded the examination questions 'Now' when I hadn't even begun to prepare the lectures! There was only one solution; to set the questions now, and then design the lectures around the questions (the full horror of this situation came back forcefully to me recently when I glanced at my old lecture notes and found that I could no longer make sense of any of the formulae that I used to teach).

I managed to set all the question papers on the printer's schedule, only to find that I was having to give lectures faster than I could prepare them, doing research and learning about some subjects only a few days ahead of my students. It was a period of the most intense intellectual effort of my life, but the adrenalin was pumping high and I revelled in the challenge (almost) every minute.

It was typical of Prof Evans that he chose this critical period to enrol me once more for a PhD degree; of course, I gave him no inkling of my problems, which would never have happened to him. It was just as well, for my harassed mental state, that other factors conspired against my actually doing any research work, such as shortages of cement and steel, which could not be bought without licences. The Ministry was more interested in providing materials for rebuilding property destroyed during the War than in giving licences to make beams that we would promptly break in our testing machines. If it had not been for the surreptitious generosity of his friends in the construction industry, who could spare a few bags of cement, I doubt if Prof would have got any experiments done at all.

Even though the war had finished, life continued to be difficult. Britain was bankrupt and could not afford to import many things that even in the straitened conditions of the 1930s had been taken for granted. Food, petrol, and clothing were still rationed, but there was also a dearth of scientific equipment, instruments and, particularly, books. I had to lend my own textbooks to the university library so my students could borrow them—they were otherwise unobtainable. I never got them back.

Telephone lines from the main switchboard were limited; our department had only one, and long-distance calls had to be booked hours in advance. I didn't receive many incoming calls, but when they did come, I was invariably in the ground floor laboratory and had to run up four flights of stairs to get to the phone. It kept me fit.

Finding equipment became a serious problem; our meagre stock of pre-war surveying instruments could be described as 'scrawny.' We hoped to buy some at a sale at Farnborough aerodrome, which involved an all-night journey to London, a whole day at the airfield, and another eight hours overnight getting back to Leeds—with no sleeping accommodation or meals on the trains in those days. However, all we could get was an RAF aerial survey camera, and when I arrived at work the next day at 9.00 am, after snatching two hours' sleep, Prof queried why I was so late; after two sleepless nights *he* hadn't thought it worth going to bed when there was so much to do.

A short time later, I was dispatched to the London docks to inspect a testing machine before anyone else could buy it, but it proved to be nailed up in a crate on top of a monumental pile originally intended to be shipped to Russia under the lend-lease programme, so we had to buy it unseen. Some weeks later, after we had assembled and installed it, complete with Russian markings, I was summoned to take an urgent trunk call from London. While I recovered my breath from dashing up all those stairs, the caller said that 'they were sorry to disappoint us, but we couldn't have the Russian machine because it had been sold to the Civil Engineering Department at the University of Leeds'.

At about this time, Prof secured for me a pokey little room on the staircase that had originally been designed to house four water closets. I was lucky to have my very own room, too small to be shared by anyone else. The Music Department was at the top of our staircase, and I was entertained when James Brown and John Boorman were practising Bach's St Anthony chorale on two pianos—a piece I loved. But atonal composers such as Alban Berg and Schoenberg sent me scurrying down to the laboratory.

This post-war chaos created in me a feeling of insecurity, and this was unexpectedly reflected at our annual surveying course in Whitby. I had arranged for our students to run a series of levels along a stretch of road from one benchmark to another (benchmarks are scattered all over the British Isles, their height above average sea level being shown on Ordnance Survey maps. They are a permanent base of reference for all building work, carved in stone or set in concrete on a very firm base and regarded as absolutely reliable).

The only student who did not achieve results within an acceptable margin of accuracy was Chapman, who couldn't make his figures agree at all. I couldn't find any fault in his calculations, so I dug out our most reliable level and together we surveyed the road from one benchmark to the other and then back again. His readings were 'spot on'—the levels shown on the Ordnance Survey maps were definitely wrong! It was unbelievable, for builders and engineers depend upon benchmarks that are revered for their accuracy. Every other student in the class had 'cooked his books' to fit the numbers shown on the map. It was a pity that we never held the course again in Whitby, for this situation would have been a unique test of skill and honesty, at least until the students learned the truth.

Chapman was a charming young man, and so brilliant that, after seeing his exam results, the Professors of the Mathematics and Physics departments both pestered him to change to a Pure Science 'suitable to his special talents' (they were known to regard all engineering as 'mere technology' and unworthy of University status). But he modestly explained that he had seen so much suffering in China that he was determined to qualify as a civil engineer and return there to improve living conditions, by means of irrigation, water purification, sewage disposal, and so on. No flattering inducement would sway him, and with his First-Class Honours degree he returned to a country that was dominated by innumerable warlords and was then in the throes of a civil war between Chiang Kai Shek and the Mao Tse-Tung communists.

We never heard what became of him, but we became anxious when Dr Sutherland told us how, as a medical advisor reporting on public health in the 1930s, he had found in China a culture that was mediaeval. A Warlord would take over a town and plunder it mercilessly. When another bandit threatened to attack the town, the first would offer to leave quietly, on payment of a departure fee, 'to avoid unnecessary bloodshed'. The incoming gang would repeat the whole process. On one occasion, as Sutherland was traveling along a very rough country road, he met one of these Warlords coming towards him in a convoy of vintage vehicles. Sutherland had an armed bodyguard, and should have felt secure, but he was very nervous because every bandit kept his ancient gun trained on him as they passed by and the jolting of their elderly vehicles on the bumpy road might easily have jerked one of their trigger fingers at any time.

The next element of chaos to arise was a series of sketch plans for a new Engineering building submitted for our approval. Prof and I had been waiting for these with keen anticipation, for they were desperately needed, but we were appalled when we reviewed them. They were in a style then contemptuously derided as 'Great West Road Architecture', with 'A Queen Anne frontage and a Mary Anne backside'. Much more seriously, the architect had not troubled to visit any of the departments to examine their needs. Had he done so, he would have learned that mechanical engineers required substantial concrete foundations for heavy engines driven by petrol, diesel, and steam, pumping machinery,

Chapter 4: 1945–1978 University Life

turbines, and so on and civil engineers needed heavy foundations for testing machines to bend and break huge concrete beams. When we saw Mr Lodge's drawings showing such testing machines and reciprocating engines located on upper floors where their vibration would have had a catastrophic effect on the building, Prof exploded with disgust, and all the professors agreed that he must be sacked. I was asked to make a sketch design to illustrate what we really needed.

The position was delicate. I had passed all of the final examinations and had only to submit my thesis and sit the next exam on Professional Practice to register as an architect. I felt that it was inappropriate for a student to show an eminent Fellow of the RIBA[1] how he should design our building, but the departments were desperate and insisted that they couldn't afford to waste time dealing with someone in whom they had lost all faith. So I set to work furiously, and by the autumn of 1947 had produced a complete set of plans and elevations that I hoped would blend with the original Lanchester and Lodge buildings. These provided ground floor laboratories for the heavy equipment, surmounted by a multi-storey spine containing small research rooms, staff rooms and drawing offices—flexibility being the keynote of the whole design. All four professors were delighted with my efforts.

The Vice Chancellor[2], Charles Morris, then arranged a meeting for the professors to explain to Mr Lodge the shortcomings of his design. This was my first sight of Mr Lodge. He had a distinguished appearance with snowy white hair, just long enough to be 'artistic', and had elegant manners. He was a charming man and, eventually, I came to like him very much. But being unaccustomed to any serious criticism of his work, he was visibly shaken, and was thoroughly taken aback when shown the plans that I had prepared, so our working relationship got off to a shaky start.

The fact that a reputable firm should have presented such incompetent sketch proposals for a multi-million-pound project, came as a great shock to me (Lodge's drawings were amateurish and some let-

---

[1] Royal Institute of British Architects
[2] The Vice Chancellor (known as 'the VC') was the Administrative head of the university – the position of Chancellor being merely honorary and usually filled by a 'Very Important Person' who would bring glamour to official occasions such as degree ceremonies, and the opening of new buildings.

tering so illegible that we had to guess what some words were). However, his chief assistant, Allan Johnson, who had been at the meeting, was enthusiastic about my proposals. He was the executive architect in charge of all the firm's projects in Leeds and seized this opportunity to produce a design based on mine but developing and improving my ideas. The departments were delighted with his drawings, and as Lodge happened to be in hospital at the time, they were rushed through the appropriate Committees, Senate, and Council in record time.

I now turned to finish my thesis on pre-stressed concrete bridges working late into the night and all weekends to produce, very laboriously, three copies of typewritten text, using inexperienced clumsy fingers on an old pre-war typewriter, and two sheets of carbon paper. It was slow, because every mistype (and there were plenty of those) had to be carefully erased, one by one, on each of the copies. Every one of the illustrations, forty-two in all, had to be reproduced exactly three times over. This amounted to a total of 126 diagrams in pen, pencil, and watercolour wash (in those days the only other means of duplicating diagrams mechanically was to have used a Roneo duplicator that produced thick, clumsy lines, which the girls in the office hated because the ink stained their fingers and got on their clothes). As a precaution, I loaded the thesis with plenty of calculations on stress, comparisons of depth/span ratios and so on, which I hoped would impress examiners who might be architects, and not engineers. I was rewarded with a 'Distinction'. The exam on Professional Practice was passed in 1952 so, after nine years' effort, I could call myself an architect.

The VC then suggested that I might join the Administration 'to advise him on all matters to do with architectural design and building work'. This sounded attractive, but the invitation put me in a quandary. Abandoning my PhD (once again) did not worry me half as much as the possible effect that such a post might have on the series of new lectures I had just initiated, which were very close to my heart.

I had been appalled to see the hideous road bridges that had just been built over our very first motorway, the M1, which should have been a prestigious image of post-war Britain. An eminent pre-war Engineer, Sir Owen Williams, who disdained pre-cast and pre-stressed concrete, insisted that the concrete should be cast 'in situ'. However, construction of formwork for 'in situ' concrete is slow and cumbersome; the consequent delays to the progress of the road were becoming serious, and the bridges, which were embarrassingly ugly, were universally condemned. When I spoke to Prof disparagingly about these clumsy designs, comparing them with the beautiful bridges to be seen in Europe, he simply said, 'Well you'd better do something about it'. Accordingly, in 1947 I devised a course on 'engineering aesthetics' designed to enable our students to create bridges and large structures without recourse to advice from architects. This covered the fundamental principles of aesthetics and included lectures on the brilliant bridges by Robert Maillart in Switzerland, the exciting thin concrete shell structures that were currently being built in Mexico by Felix Candela, and works in Italy by Luigi Nervi.

An Owen Williams bridge contrasted with the 'Needle Eye' bridge, influenced by Robert Maillart.

I accepted the VC's invitation, but only on condition that I could continue to develop the course, to be called 'Architectural Engineering'. I thoroughly enjoyed giving these lectures, and the course developed nicely, with interesting 'drawing office studies', but after ten years, by 1957, I had become so heavily involved in the university building programme that Prof had to appoint an architect, Bill Houghton Evans, to take it on full time.

One of my brightest and most enthusiastic students, Ted Happold, delighted me by spending his long vacation working in Alva Aalto's office in Finland. He later joined Ove Arup and worked on the Sydney Opera House and the Pompidou Centre and became President of the Institute of Structural Engineers and Professor of Architectural Engineering at Bath University. Ted was knighted in 1994, and I am very proud of his success. Now, fifty years later, there are two courses at Leeds; 'Architectural Engineering' and 'Civil Engineering with Architecture' so, all in all, I am well pleased and very satisfied. It was especially gratifying to learn that the prestigious concert stage for the Queen's 2012 Jubilee concert in front of Buckingham Palace had been designed by Neil Thomas BSc Architectural Engineering, Leeds, 1980.

When I started my lecturing career, I was living with my parents in Bradford. Bus journeys proved to be a waste of time I could ill afford, and since there was a waiting time of more than twelve months for any new car, I bought a little pre-war 'Flying Standard' for my daily commute to Leeds. Its eight-horse power engine achieved a maximum speed of 45 mph.

All went well during the summer, but Yorkshire winters with snow, ice, and fog were formidable and worrying when I had to reach the university to give a 9.00 am lecture. These early lectures were unavoidable, for the only way I could produce the complicated astronomical drawings required for the surveying course was to draw them on the blackboard the night before, with a prominent notice threatening dire consequences for anyone touching the blackboard before 9.00 am. After one or two 'hairy' early morning journeys over ice and snow (when I discovered that even a cautious application of the clutch on an

Chapter 4: 1945–1978 University Life

My aunts, my 'Standard 8', and me.

icy road could result in the car slewing round to face oncoming traffic), I moved to Devonshire Hall. It was a good decision; I not only saved precious time but developed invaluable friendships that sustained me for many years, and my room had enough space for a large drafting machine.

One incident was unforgettable. I had been working on my RIBA thesis all through the Saturday night and was at the formal midday Sunday Lunch when James Brown remarked that he hadn't expected to see me—wasn't I supposed to be dining as a guest at a Women's Hall of Residence? It was a nasty shock. I was in deep trouble and hastened to Weetwood Hall to present my explanation and apologies with the largest bunch of flowers I could find, assuring myself that there must have been several other guests and my absence might have gone unnoticed. But, to no avail; the Warden received me icily and made it clear that I was to have been her personal guest, and the vacant chair on her right hand had been an embarrassment throughout the meal. She insisted that there was nothing I could do to make amends, and she took great care to avoid me from then on. I regretted giving offence, for I liked Miss Carey, but I was far too busy to go to any social affairs that I could avoid. On further reflection, it dawned on me that I might even benefit from missing this event, for I realised that my perfidy would be noised abroad, especially among the Women wardens. They would never know with what relief I welcomed the prospect of not being invited to any of their halls again!

James Brown and I regularly played squash and sat together at the high table whenever possible, for James was anxious to avoid the Warden, who he disliked intensely. Commander Evans was a distinguished Naval Officer, awarded the Albert Medal when he rescued a seaman from burning oil at the risk of his own life. It was the highest naval decoration, and equivalent to the Victoria Cross, but our genuine respect for him could not alter the fact that he seemed to be out of place in our academic environment. He spoke with clipped sentences and ran the hall as if it were a warship, striking a large ship's bell at mealtimes and treating the students like seamen who needed to be disciplined. This might have been acceptable if they had been young boys fresh from school, but many were mature ex-servicemen with distinguished war records, and his attitude didn't go down well with them. Members of the Senior Common Room included professors and senior academic staff who had resided at Devonshire Hall long before Evans had been appointed. They all outranked the Warden and were resentful when he treated them like his juniors. They regularly attended student events, because getting to know students was the whole point of living in halls. So, when Evans developed a habit of pressing them to attend every function, as if it were a favour to him, they found it very irritating.

It was at the Annual Devonshire Ball that the Warden's ambitious entertaining reached its apogee, when he invited wardens of other halls and senior professors to accept his hospitality. As usual, he asked the Senior Common Room to help in his entertaining. Although he must have known that most of us detested dancing, he probably never imagined that his singing of songs from Gilbert and Sullivan bored us to death. In particular, James was sick and tired of having to accompany him, for he quite rightly considered it demeaning for a Senior Lecturer in Music to have to play such trivia as comic opera. So James decided to rectify matters, and for this special occasion offered to bring in a second grand piano and to perform a duet with his friend John Boorman, who was so gifted that he might well have become a professional concert pianist, rather than a lecturer in Spanish. James was a gentle, quiet man, with a hesitant speech bordering on a stutter that gave him something of an aristocratic mien, so it was a shock when I discovered what he had planned. The only warning I had was when he whispered to me at supper that if I intended to join the guests in the Warden's lounge, I would be advised to remain at the back of the room, near the doorway.

When all the guests were assembled, the warden, obviously very pleased with the impression that two grand pianos had made, introduced James and John, and the performance started with a vengeance. It was an atonal piece by Schoenberg, and the discordant cacophony was pounded out with such enthusiasm that the room seemed to shake. Those of us at the back were glad to be able to retreat into the corridor, to conceal our mirth as much as to get away from the dreadful chords. It went on—and on—and on, until we became very sorry for the audience. When it ended, there was a shocked silence. James had been clever, for one had to admit that such a recital was eminently suitable for a distinguished academic gathering, and some people pretended that they had actually enjoyed it. The Warden withstood the ordeal like a true naval officer, making it clear that he had arranged it all, and accepted his guests' congratulations gracefully. But it did the trick, and James was never required to perform again. He became the toast of the Senior Common Room.

In the meantime I had developed a strong friendship with Dr Len Watkinson, of Colour Chemistry, who introduced me to the delights of Scottish dancing; intricate and invigorating, and a wonderful relief from daily stress. Len had been a keen scout, and when a group of students approached me to form a University Rover Crew, I agreed on condition that he would act as Assistant Rover Scout Leader. We went on a Naval sailing course at Gare Loch, near Glasgow, which enabled us to teach dinghy sailing to the Rovers on a boathouse on Lake Windermere owned by the Boy Scouts Association.

When Ronnie Morgan, Head of the Physical Education Department, and I visited the Lake District to survey and rent a derelict barn in Dunnerdale, the Rovers formed working parties to convert it into a mountain hut. This gradually developed into an important facility for several student societies. Our climbing expeditions took place in autumn or winter when bogs, snow, rain, and mist created hazards, resulting in many anxious moments for me. But it kept me fit and I learned the exhilaration of abseiling down a vertical cliff face on a single rope (not, as in nowadays, with a safety helmet and several ropes, blocks and tackle).

So the four years I spent at Devonshire Hall were not uneventful and gave me my first glimpse of social life, although Mother was horrified when I let it slip that it was customary for members of the high table to retire to the Senior Common Room after formal dinner to take a glass of port or Madeira wine with their coffee—it was obvious to her that I had gone to live in a 'sink of iniquity'.

It was while living at Devonshire Hall, that I had accepted the VC's invitation 'to advise him on architectural matters', under the impression that this was to be a temporary state of affairs, and not a permanent change in my career. But it transpired that I was to replace his Assistant Secretary, Mr J J Ilett, who was about to retire, and that I would be known as the University Planning Officer, responsible for all buildings and their maintenance, cleaning staff, porters, security, night watchmen, gardeners, groundsmen, telephone operators, joiners, plumbers, electricians, boilermen, catering services, and the management of the Halls of Residence. In addition, I was to take over the management and control of all present and future building contracts.

So Prof Evans appointed staff to take over my lectures, apart from those on Architectural Engineering, and it was arranged that I should understudy Ilett for the last twelve months of his tenure to learn all about the university's administration, the major in-hand building projects, and to prepare instructions for the next major works to be planned. Ilett loved to reminisce, and it would have been pleasant to listen to him talk if his information had not been so alarming for me. Of course, *he* was not at all worried, because he was looking forward to his retirement, and was content to lean back in his chair and puff on his noxious pipe as if to avoid any contact with papers that were accumulating on his desk, presenting a scene of utter chaos. As he did all his own filing, his secretary had no idea how to find anything, and when

I pointed to some unopened letters on his desk he proudly expounded his policy of 'masterly inactivity'. This held that 'if a problem is left alone long enough it will usually solve itself'. He explained that he knew what the letters were likely to be about, so there was no hurry to open them. I didn't applaud this theory, for I realised that I would be the one who would have to deal with any problems that hadn't solved themselves. It was clear that there were plenty of those, the most serious and immediate one concerning the Library, so I asked Ilett to outline the historical details for me.

Lord Brotherton had made a bequest so generous that this was the finest building on the campus, having marble columns from Sweden and only the very best quality hardwoods, bronze fittings, and craftsmanship that made it the 'jewel in the university's crown'. The woodblock floor of the main reading room had been floated on rubber to reduce impact noise, and the Librarian had insisted that no water should be installed anywhere near the reading room or book stacks, so water damage to his valuable collections could not possibly occur. Washrooms were therefore situated in a remote basement area. Heating by hot-water radiators was ruled out; electric heating was considered to be too great a fire risk, and so it was decided to use a 'Plenum' system using circulated warm air. As the Leeds atmosphere at that time was laden with soot, Dr Faber designed an elaborate filtration system to clean it using sprays of hot water (known as a scrubbing plant), followed by refrigeration to remove excess moisture, and finally heating it up again to warm the building. To save money, most of the cleaned air was to be re-circulated, with 'fresh air' introduced only as required.

Unfortunately this elegant scheme was doomed, for it failed to gain the approval of the university committee that included eminent professors of science and technology, experienced in the examination and criticism of documents of all sorts. When they reviewed Dr Faber's proposals and asked, 'Did he really think that the climate in Leeds would ever justify a cooling system?' his confidence wavered and he replied that he 'supposed it would not'. The ramifications of his pathetically inadequate reply were to prove catastrophic, for the committee were completely misled and promptly decided to save a lot of money by eliminating the very expensive refrigeration plant (of course, he should have explained that the refrigeration plant had nothing at all to do with cooling air in the summer, whereas it was absolutely vital to dry the air after the washing process had removed the dirt. I reflected that this incompetence of our consulting engineer was on a par with our consulting architect, and the sooner we got rid of them both, the better).

The scale of the catastrophe became evident at the very beginning of the cold season when the heating system was first put into operation. The injection of warm, wet air into the cold building produced condensation on a formidable scale; all the walls were soon streaming with moisture and all the books and documents were in such a state as to make the horrified Librarian and his staff despair. And this, after all the Librarian's insistence on the avoidance of damage by water—it was unbelievable! The labour involved in examining and drying every one of the thousands of books and archival material was prodigious.

This proved to be only the start of the Library's troubles. Once the scrubbing plant was shut down, there was no means of cleaning the filthy Leeds air and the dirt became intolerable. In desperation, the maintenance staff inserted some canvas filter bags, the best they could find in wartime, into the main duct, only to discover that the feeble fans were incapable of forcing the air through them because they had been designed to handle a smooth flow of uninterrupted air. So, to get some heat into the building, some of the bags had been removed, with the result that most of the air was now bypassing the filters entirely and the Library was being supplied with warm but very dirty air. On cold days, when this was re-circulated and re-heated, it became so stale and dry that staff and students were complaining about sore throats and headaches, and some were becoming ill. I had the greatest sympathy for the maintenance staff who had struggled with this situation and who must have been very relieved to hear that the Vice Chancellor had referred this problem to the new Planning Officer with instructions to solve it at all costs.

When I learned of the appalling sequence of events, I resolved that, at the first opportunity, we must appoint a qualified mechanical engineer to vet any more proposals from such eminent consultants who, to my amazement, still retained their appointment. They could only suggest that a refrigeration plant would solve the problem. Unfortunately, that 'stable door' had closed long ago since the architects had modified the concrete ducts, leaving no room to spare in the foundations and sub-basement. At this stage, their proposal would have required drastic structural alterations inside our precious new Library, and, beyond that, the refrigeration plant itself would have to be housed in a new, purpose-built structure in the courtyard outside. This would be an excrescence offensive to the aesthetics of the area, completely out of keeping with the ethos behind the courtyard space. Apart from all that, I was very concerned about the ridiculous waste of money involved in a convoluted system that would warm gallons of water to wash the air, and then freeze it to get rid of the moisture, and then heat it all again to warm the building.

While searching for a more efficient type of filter and more powerful fans, I heard, with great excitement, that the Americans were developing an 'electro-static' filter that had literally no air resistance and was incredibly efficient in removing even the smallest particles of dirt. It seemed that we should only have to replace the huge canvas filter bags with a screen of fine wires. I could hardly credit it, especially when the VC pointed out that, as there were not any of these marvellous things in the UK, I had better go to New York to inspect one. I breathed 'God bless America', especially when I thought about the goodies that were available at that time in the USA, such as chocolate, nylon stockings, and so on. Professor Denny had told us about his wealthy aunt who had booked a berth on the first liner to cross the Atlantic after the war and had embarked on a spending spree in New York. On her return to Southampton the Customs Officer asked if she had anything to declare? 'Oh yes!' she exclaimed, 'I've had a simply wonderful time—just look at these lampshades!' She began to tear off the paper from the largest of the numerous parcels she had with her. The officer must have thought she was slightly mad, for he asked her to move on quickly to make way for the other passengers who were queuing behind her. So Denny's aunt passed innocently through customs without paying any duty on all the cigarettes, nylons, silks, watches and jewellery that she had bought as presents for the whole family.

Such excitement and luxuries were not for me; I already had plenty to handle with the new job, to say nothing of my Architectural Engineering teaching that no one else could handle. So the VC agreed to take a gamble; we managed to get an import licence, scrapped the useless fabric filters, and bought what I believe was the first electrostatic filter of its kind to be installed in the UK. The results were astounding; visitors besieged us, and we adopted electrostatic filters for all our future needs. To me, the whole experience was something of a miracle.

In those days, the late 1940s, to spend a day in the Brotherton Library, with its clean warm air, was a real luxury. The grimy air elsewhere—in the old College buildings and outside—made everything filthy. You could even taste the soot. Shirts had removable collars that had to be changed every day, and the traditional white lace window curtains that were de regle, even in the poorest houses, had to be washed once a week. During fog, things were even worse; newspapers reporting on one wedding during which the bridesmaid's car got lost in the fog on its way to the church and the bride's wedding dress that had started off pure white ended up sooty black.

Change came from an unexpected source, with a flood of Pakistani immigrants coming to work in textile industries in Bradford and Leeds. To the astonishment of their Yorkshire neighbours, these incomers began to wash the *outside* of their smoke begrimed houses, and the beautiful natural colours of the local stone were revealed—cream, soft reds, and greys. As this practice continued, whole streets of what had been dreary blackened housing were transformed into a cheerful environment with quiet pastel colours providing a fitting background for the brightly coloured saris of the womenfolk.

The Government then introduced the Clean Air Act, which began to reduce the amount of soot in the air. Local authorities were stimulated to follow the example set by the Pakistanis and began to wash the outside of their public buildings, encouraged by scientists who advised that this would save money by preventing further erosion from the acid soot. The effect of washing the cream-coloured Portland Stone of the Leeds Civic Hall was dramatic, and, in due course, the university followed suit, for there is nothing that a Yorkshireman likes better than saving money. However, I doubt if many gave the Pakistanis any credit for their initiative.

Chapter 4: 1945–1978 University Life

Ilett went on to tell me how progress on the new Parkinson Building had been seriously held up in the 1930s by a lack of drawings, for Lanchester was far more interested in winning more architectural competitions than in the dreary work of producing working drawings for designs he had already won. The university, in frustration, finally decided to sack him, but in a desperate effort to save the firm, his partner, T A Lodge, begged the Vice Chancellor to appoint him personally as an architect, as distinct from the firm of Lanchester and Lodge, promising to produce the necessary drawings forthwith and to see the job through efficiently. This he did; the shell of the building (apart from the tower) was completed just in time to be made sufficiently watertight for use during the war as a government food store. Work was resumed after the war, but it was hindered by shortages of skilled labour and materials, many of which were rationed (although our sub-contractor, Nicholson, had hoarded enough of their precious stock of walnut to panel the walls of the splendid council chamber). However, these difficulties were trivial compared with the problems created by the university itself.

The original specification had called for a science building with large teaching rooms and laboratories, and that is exactly what had been built, but when Frank Parkinson, who had donated most of the money for the building that was now to bear his name, expressed his wish that the Council Chamber and the Vice Chancellor's offices should be housed in it near the main entrance, the architect was instructed to modify the plans accordingly. I am sure that no one could have foreseen the ramifications that this decision would entail; they develop remorselessly as one problem after another evolved in sequence first intimation of trouble came when the VC pointed out to although his room was dignified and spacious, he could on the surrounding grounds by standing on a chair, for the were so high; they had been designed to allow laboratory shelving to be placed underneath them. So it was deci wooden floor some twenty inches above the concret until it we realised that for safety and convenience for the Registrar and all his clerical staff should Even that didn't deter us, until the builder s openings to suit the floors. The noise, dust, and delays were beyond belief, because Sir Edwin Airie had evidently intended that his building would 'last a thousand years' and every brick was of 'engineering standard'—incredibly tough, and the mortar was the strongest it could possibly be. Even the partition walls were made of solid twenty-one-inch-thick brickwork.

It was to get much worse, for senators visiting the site began to wonder whether their voices would ever be heard in the new Council Chamber, which was located above the main entrance facing Woodhouse Lane. The screeching of tramcar wheels, and the grinding of gears of heavy lorries crawling uphill, easily penetrated the single glazing of the windows on that façade, which was now to accommodate all the administrative offices under the control of the VC. They had no hesitation in deciding to change to double-glazing, although this now meant that air conditioning must be installed in the whole building. The insertion of ducts required the breaking out of hundreds of openings through concrete panels and brick walls built to the specification of Edwin Airie. Thank goodness I had no responsibility for the finances of that contract and was happy that we could use the electrostatic filters in the artificial ventilation system.

As the monumental aspect of the building grew in significance, senators began to regret their lack of foresight; no one had thought to incorporate into the foundations the traditional 'time capsule' to be discovered by future generations if ever the building's remains were to come the subject of archaeological exploration. They decided that the best thing would be to encase a sealed casket in the summit of the O with such ceremony as could be contrived by the new Planning It with the Clerk of Works and the master mason in attendance. this flimsy temporary scaffolding, and my vivid recollection of ladder is how white my knuckles became, clinging to a shaky the Park ddy height; how glad I was to descend at last back to As a the clock t ought, the Senate decided to install bells in arrangement enny, professor of music, composed suitable battle against nd quarters, but had to fight a prolonged rified him by proposing to save money by using electronic chimes. The argument was only resolved when James found a donor willing to pay for traditional bells. Getting them into the tower was left to me, and I was very grateful for the quality of the builders work, because their weight and vibration might have had disastrous consequences in a weaker structure. (In 2015, a pair of peregrine falcons made this spot their home; it seems that the bells, that must have been deafening at close quarters, didn't worry them at all.)

It soon became evident that I was expected to organise the re-arrangement of accommodation for some twenty-seven departments who benefited from the newly acquired space in the building (I learned much later, that Charles Morris had commented to other vice-chancellors, 'What a relief it had been, to have had a planning officer to deal with the onerous job of dealing with so many heads of departments and reaching agreement, without him having to lift a finger').

In fact the manipulation of space was like solving a jigsaw puzzle, and I enjoyed the exercise because it was a wonderful way to meet so many professors and staff, many of whom became close friends, and I did not find it 'onerous' because most of those involved were very pleased to benefit from the proposals I put to them. The irony was that the only ones with a significant complaint were the mathematicians who had been allocated the most prestigious accommodation on the top floor of the Parkinson; they soon discovered that the only lavatories in the six-storey building were in the basement. Unfortunately the lavatory block, originally planned for the south end, and the elevators that would have served all floors, had been omitted from the contract to keep within the available funds. So, the mathematicians had to go down twelve flights of stairs to use the student lavatories in the basement and then climb back again. Their apoplectic outrage was understandable, and even when I scraped together enough money to install a cheap lift, it was slow, and so inferior that it kept breaking down; a contract to build the lavatory block and install two decent lifts became my top priority.

In the meantime, the formal opening of the Parkinson became a gala affair, with a degree ceremony at which 'Tommy' Lodge was given an Honorary Doctorate, and a full-dress ball was held in the Parkinson Court—the large atrium on the ground floor. Members of staff

The Parkinson Building

dancing in their colourful hoods and gowns threatened to eclipse their ladies' evening dresses—an unforgettable occasion, marred only by the danger of tripping over the bases of the columns that supported the tower and projected into the Court, which persuaded us never to repeat this event.

In parallel with the Parkinson problems, I had been told to tackle our most urgent need—the demand for more catering facilities. Before the war, some 70 percent of students had lived at home, brought sandwiches for lunch, or had a snack in the bar in the Union Building, leaving 30 percent to lunch in the refectory, an ancient malt house, where dampened malt used to be spread out over the floor to germinate, and whose open timber roof trusses still exuded the sweet odour of malt. It had two modest dining rooms, one for men, where women were rarely seen, and one for women, where men were never seen. In the 1930s, when money was really tight, luncheon vouchers for the refectory were offered at 'Bazaar Days', which were held at the beginning of each session, to register students for the courses they had elected to take. Academic staff were stationed round three sides of the Great Hall, and students queued in long lines in every direction. The bustle and hubbub was intoxicating, especially to freshers who were bewildered by the whole affair. Being stuck in a slow-moving queue made them easy prey to older students who pestered them to join the OTC, air squadron, one of the innumerable societies, and to extract a subscription from them before they had a chance to think twice.

When the war came along, free breakfasts were provided for everybody on fire-watching duty. We slept, fully clothed, on camp beds in the Joint Common Room under the Great Hall. In the winter of 1939–1940, I was very glad of this arrangement, for I was working every evening in the Brotherton Library swotting for my finals, or in the structures laboratory on my research project.

After the war, these very primitive facilities were overwhelmed by ex-servicemen returning to the university, and government grants becoming available, resulting in only 30 percent of students now living at home, leaving 70 percent demanding lunches and evening meals. A temporary prefabricated concrete cafeteria on the tennis court had done very little to alleviate the pressure on the old refectory in which everything remained exactly the same, except for the very long queues. The only other catering unit was 'Staff House', carved out of properties in Beech Grove Terrace. Dining rooms there were small, having tables for four, with an atmosphere of exclusive intimacy that appealed to people who sought privacy and an opportunity to gossip, rather than to expose themselves to the raucous freedom of the Refectory. Staff House also had a miniscule common room with newspapers and periodicals.

It was then, as a very junior member of staff, I discovered the Staff Dining Room in the Refectory, with its own discrete entrance. The meals were the same as those for the students, but instead of long tables, there were small ones surrounding one enormous central table seating up to eighteen, which was by far the most popular place to dine. For it was here that arguments developed between erudite professors, and when their exchanges erupted into good-natured badinage the witticisms could be hilariously entertaining. There was keen competition to sit with them, and this large communal table was a feature I felt must be provided in the new building that was now my first priority. The Vice-Chancellor forecast that student numbers would rise to

5,000 within ten years, with correspondingly more academic staff who could afford better standards of catering, and were demanding evening meals as well as lunches, so this gave me a basis on which to base my schedule of accommodation, which called for a building to serve more than 3,000 lunches a day.

There was only one possible, but very restricted, site available between the Union Building, the tiny gymnasium, and Lyddon Hall, and I must admit that Lanchester and Lodge did well to fit such a large building onto it. Work on the foundations was well in hand when my friend Ronnie Morgan, Warden of Lyddon Hall, burst into my room shouting, 'Those blasted contractors for the Refectory are digging a damn great hole that's undermining our Hall and our kitchen is about to fall into it!' A frantic call to London elicited a very calm, relaxed response from Lodge to the effect that he 'had assumed that Lyddon would be demolished'. *Demolished*—a thriving men's Hall with sixty residents and a long and honourable history, and he hadn't thought of mentioning this to anyone? Of course the kitchen did not fall into the hole, but it had to be taken down and the building operations were so close that the members of the hall had to put up with a temporary kitchen in their backyard, accompanied by intolerable noise and dirt from building work for two years. After that, we built for them a very nice new kitchen.

The new Refectory had a magnificent dining room seating 1,200, which was also used as a ballroom. It adjoined a special dining room with waitress service for 120, open to everybody, so that students could entertain their parents and friends. This proved to be a resounding success, for, until then, there had been nowhere for staff and students to mix socially, and parents and visitors could only be entertained by being taken downtown to some restaurant in Leeds. The Senior Common Room filled the top floor with a large dining room, three small ones for entertaining visitors, and unheard-of luxuries such as common rooms, library, card room, billiards room, and so on. The manageress, Miss Rhodes, even had a separate office for her secretary, to the envy of many academic staff, who, at that time, were being forced to share small staff rooms.

The basement had a coffee bar, a cafeteria, and a bar from which alcoholic drinks were served. Strong objections by some Senators to an alcoholic bar were stifled when it was agreed that it would be open to members of staff as well as students. As all the rooms in the building were serviced by lifts from the main kitchen, I was anxious about it functioning smoothly, but to my delight, Miss Brownlie, with Miss Rhodes as her deputy, transferred successfully, from the simple Old Refectory and Staff House, to this megastructure with its incredibly complex services—apparently without 'turning a hair'. The only serious problem they had faced was an unaccountable build-up of queues at peak periods. As lectures and classes finished, either at noon or 1.00 pm, queues would begin to develop, even extending outside the building, causing bitter complaints from students, especially in wet and windy weather. When they threatened to become riotous, I was told to 'do something about it!' It was a puzzling situation; it didn't make sense! Miss Brownlie, with years of experience to guide her, knew within fine limits what the demand for each item on the menu would be. For example the distribution might be 50 percent meat, 35 percent fish, 10 percent vegetarian, and 5 percent salad. The flow of food to the serveries was based upon these proportions, but the queues along the serving counters were being repeatedly held up, and no one could explain why this should be so until we asked a Behavioural Psychologist to examine the situation. He explained that Miss Brownlie's system of distributing food to the serveries was perfectly logical. It was the customers' fault that queues arose, because they were human, and fallible, and were influenced by others as they moved along the counter. Some tended to change their minds and followed the choice of the person in front of them, and the sight of several people choosing the same dish encouraged this tendency and created a surge that exhausted that particular item and stopped the queue. Any unusual or exclusive dishes were particularly susceptible to this effect. When we followed his advice and arranged a strategic back-up of these particular items the problem was solved without recourse to any of the very elaborate rearrangements I had begun to examine. To my relief the only expense was the consultant's fee.

After the debacle of Lyddon Hall, Lodge didn't often appear in Leeds, but one day he invited Dr Belton and me to tour the nearly finished building to settle the colour scheme. He ceremoniously conducted us through the building (Allan Johnson had divulged to me that 'Tommy' used ceremony to compensate for his small stature, and would never enter a committee room until everyone was seated, thus ensuring that the chairman would give him a formal welcome and arrange appropriate seating for him). He started by suggesting a pale green for the entrance hall that opened on to the main corridor, which 'might be done in a darker shade of green'. The main dining hall, he thought 'would look well in olive green', and on the lower floor 'a lighter lemon green would obviate the basement feeling'. As we sat down to discuss these elegant ideas, Joe Belton whispered, 'Bet you can't guess what colour his socks will be?' He was right! In the event, we both had too much to do elsewhere to waste time arguing about his precious greens, and we didn't raise any objections.

In the meantime, I was settling into my new post as Planning Officer in Ilett's room, directly above the Vice Chancellor's room near the main entrance to the Yorkshire College Buildings and the Great Hall, which, at the time, was the headquarters of the University Administration. As a student, I had been under the impression that the Hall Porter and the VC's Secretary actually ran the university, for they certainly dominated all aspects of student life, and Hilliard and Miss Selby were always to be found in this area. The Hall Porter Hilliard was tall, straight as a ramrod, and his military ribbons, lofty bearing, reflecting his career as a Regimental Sergeant Major, supported his lordly treatment of students and junior staff. He would condescend to treat professors as equals, but his genuine respect was reserved for members of the Royal Family. Students were in awe of him, especially when we learned that whenever he was required to escort the wife of the then VC, Lady Baillie, he would precede her through crowds of students giving her the impression that he was saying 'Make way for Lady Baillie'. But, in fact he was delighting the crowd by actually saying, Make way for the old b***h'.

I wondered how I was going to cope with him, now he was a member of my staff, but I needn't have worried—he just ignored me and went to the Bursar to voice his complaints, often threatening to resign. He did this once too often and was astounded when the Bursar accepted his resignation. Miss Selby was delighted, but things changed when the Chancellor, the Duke of Devonshire, arrived to attend a Degree

Ceremony. He was a charming old man, and had known Hilliard for many years, so as he stepped out of the car at the foot of the Parkinson steps, his first words were 'Now, Hilliard, I hear you're leavin' us—we can't have that y'know!' Miss Selby, standing nearby was appalled; as soon as he had escorted the Duke to meet the waiting dignitaries, she could not stop Hilliard from rushing off to the Bursar to withdraw his resignation on the grounds that 'The Chancellor won't allow it!' Consequently, Hilliard became even more confident in his indispensability until regulations ultimately required his retirement on the grounds of age. Everybody liked his deputy, who took over. He was courteous, efficient, and all that could be desired, except that he was not so tall and as easy to find in a crowd as Hilliard had been. So, there were odd times when we actually missed the old tyrant.

In complete contrast was Lady Ogilvie, Tutor of Women Students, who I knew from my attendance at various committees. She radiated a kind of warmth that must have been a great comfort to women students who needed her advice, and had a firm, but quiet, authority. Her suite was near the Vice-Chancellor's room at the bottom of the staircase leading to the Planning Office. Very shortly after my appointment she intercepted me and astonished me by offering the following advice; 'Mr. Wilson—your first few months will be a honeymoon period when you'll be able to do anything; do take advantage of them while you can!' So kind, and so very perceptive. I thought the world of her, and was lucky enough to find that my 'honeymoon period' was to last fifteen years—with constant support and encouragement from Charles Morris until he retired in 1963 (it ended with the coming of his successor, Sir Roger Stevens, who didn't show any interest in my work. Perhaps it's just as well, for at the time we were well launched into the Chamberlin Development Plan, and if he had disliked what I was doing, things might have been very awkward indeed).

Charles Morris had been appointed Vice Chancellor just before I arrived, and he had shocked the administration by failing to keep Miss Selby as his secretary. His decision was the subject of some comment when he appointed an attractive young lady, Miss Morrish, to this post, but the gossip stopped when he pointed out that Miss Selby had work to do that was more important than merely looking after his correspondence and promoted her to a newly created post of 'Secretary for Hospitality', responsible for entertaining distinguished visitors, and making arrangements for all social affairs and ceremonies. He was quite right. Miss Selby's new duties soon justified an assistant and clerical staff. She may have had a reputation for caustic impatience, but she evoked nothing but sympathy from me, for I had to admit that the facilities for entertaining and catering were woefully inadequate, especially regarding toilets. She confided to me that on her death, the letters 'WC' would be found engraved on her heart, for discrete access to a private toilet during the visit of any VIP was crucial, especially if that visitor was elderly, as most of them were.

So when Denis Jones remodelled the Great Hall, we planned a rest room behind the stage for the exclusive use of the Princess Royal, who had succeeded the old Duke of Devonshire as Chancellor. I had never seen Miss Selby so radiantly happy as she was on the first Honorary Degree Day to be held in the renovated Great Hall, with its new organ, new stained-glass windows, new curtains and gallery, and above all, a beautiful new lavatory behind the scenes! Her Royal Highness was the first member of the Royal Family to become Chancellor of a British university. She was the only daughter of King George V, and it was common gossip that her marriage to the Earl of Harewood was not a very happy one. The Chancellorship gave her freedom to act independently, and it was evident that she enjoyed meeting staff and students and officiating at university functions. Whether her happiness influenced the Royal Family or not, it became fashionable for other Royals to accept similar appointments at other universities. In due time, she was succeeded by HRH the Duchess of Kent, a member of the Worsley family of Hovingham, near York, who was very young—about ten years younger than me, and, being attractive and charming, her visits brightened up the whole university. She evidently relished her association with both staff and students, even attending the Union Ball to the special delight of those students lucky enough to dance with her.

I had not been in my new post for more than a week or two when the Bursar, a very large man, reputed to have a violent temper, came thundering down the creaking wooden floors of our ancient corridor, burst into Ilett's room and demanded some papers that he required for Senate that afternoon. Ilett shrank back in his chair and muttered that he hadn't seen them. It was clear that Mr Brown had lost all patience with Ilett and his relaxed demeanour and was very worried about having to face Senate without having the papers. I sympathised with him, so I smiled and said that I would find them. It was a gamble, but I eventually found them in one of the unopened envelopes on Ilett's desk. After that, Mr Brown was particularly kind to me, especially some time later when he discovered that I had been quietly circumventing some government regulations in a way to which the local Ministry of Works would have taken strong exception, but that were absolutely vital if our buildings were to make any progress. He ribbed me with the remark 'I didn't know that boy scouts did that sort of thing', and, after that, we got along splendidly.

Visits from the Duchess of Kent brightened up the whole university.

When I learned that the university was governed through committees, I was appalled to find that I was to be responsible for managing several, including the important House and Estates Committee, its several sub-committees, and the Students' Union Building Committee. The time and labour involved were unbelievable; memoranda had to be written for most items on the agenda, my 'notes for the chairman' had to be reviewed by the Bursar and then by the VC, and finally, the minutes had to be approved by them. As the work expanded, I appointed a full time administrative assistant, Louise Friend to do this work, and her expertise proved invaluable to me.

Miss Friend had been secretary to Mr Wardley, the Bradford City Engineer who had been in 'deep water' when heavy rains had flooded the city streets, including several pedestrian tunnels with lavatories that he had installed at considerable cost. Divers had been needed to search for anyone trapped in them. Louise stoutly defended him, but I was able to tease her when the press published his plan for an elevated road circling the city, for this had only one point of access; anyone using it could only encircle Bradford and then get off at the same point that he got on. It was obviously a premature press release and must have been very embarrassing for him, but he had a good sense of humour and related how the streets of Leeds made sense—'Kirkgate leads to the church, Briggate to the bridge, and Eastgate and Westgate were obvious, but he did take exception to 'Swinegate' when told that it led to Bradford.

The Chairman of the House and Estates Committee was Brigadier Tetley; a charming man who owned Tetley's brewery and whose family had been generous to the university over many years. He conducted meetings decorously, reading out my notes verbatim, so I had to be very careful about every word I put into his mouth, and spent long sessions with him beforehand going over every item in detail. The VC usually handled questions during a meeting, and I did not speak unless a technical point was raised, so meetings were boring and I frequently drafted the minutes in advance (successive chairmen were not always so discreet. Dr Dower was competent and enthusiastic but tended to rely on his memory rather than my notes and was so free with his opinions that sometimes I used to dread what he might say when it was clear that he had misunderstood the facts. Those were days that I remembered Brigadier Tetley with affection).

The government was now flinging money at universities and planning new ones in a desperate effort to accommodate servicemen and school leavers qualifying for scholarships. Our intake of students increased every year, creating serious problems for all those responsible for teaching classes, catering and student housing, and everybody turned to me for help. Subsequently, when the Surveyor of the Fabric, Mr Leybourn retired, I appointed an architect, A L Knighton, to deal with the conversion and extension of properties as fast as we could acquire them, using houses within the precinct for departments and those outside for student housing. Whenever we bought a large house with extensive grounds, we commissioned a local architect to develop it into a Hall of Residence. In this way, we found Denis Mason Jones, who was outstandingly good, designing Lupton Hall, Tetley Hall and later the extensive Bodington Hall. However, Denis couldn't cope with everything, and we had to employ other architects for the three experimental farms near Bramhope, the Extra Mural Centre at Middlesborough, and the Marine Laboratory at Robin Hood's Bay.

The financing of these minor works was significant. The VC had a brilliant mind and an extraordinary capacity to think ahead, not only to the next stage, but also to the one beyond, so that people found that he had solved their problems before they realised that they existed (he was soon invested as a Knight and later as Lord Morris of Grasmere). When he discovered that the annual funding for universities was not being fully utilised because some universities had not started their buildings in time to fit the Treasury's annual budget, he suggested that I should prepare in advance a number of projects that might 'mop up' any surplus funds that came available at short notice. In this way, by being one step ahead of other institutions, he secured thousands of extra pounds for Leeds, simultaneously authorising more assistance for me without my asking.

My first Deputy was Roy Pemberton, a wonderful colleague and lifelong friend, and I recruited Mechanical and Electrical Engineers. Assistant architects replaced Roy when he became Planning Officer at the new University of Sussex. The office steadily grew from the original three to thirteen; Denys Horner and Ron Crawford supported me loyally for many years and Robert Sladdin, a Chartered Surveyor, took up a new appointment as Director of Estates in charge of the Planning Office when the three of us retired. By that time, student numbers had grown from 3,000 to 32,000 students.

Of course, that was a long way into the future, and in the meantime we had no Development plan to guide us, so Dr Lodge was asked to prepare one. The zone allocated to the university under the City Plan included about 140 substandard houses scheduled for demolition, plus property in adjoining streets that the university was acquiring discreetly through an agent. Lodge's plan laid out a pattern of roads with buildings fronting on to them, in accordance with his philosophy that 'roads were more important than buildings'. I concluded that the winning design in the 1927 architectural competition must have been the brainchild of Lanchester, rather than Lodge, for his obsolete concept was reminiscent of the eighteenth century and was quite inappropriate for the second half of the twentieth century. I put it to the VC that we must look for a new, younger architect and planner. He agreed, but meanwhile we were stuck with the official appointment of Lanchester and Lodge for the completion of several buildings. Some senators were all for sacking the firm forthwith. I had to work hard to persuade everyone that Allan Johnson's excellent design for Engineering, now well advanced, showed that he was very competent, and that it would be financially disastrous if the existing programme were to be disrupted at this late stage. So, it was accepted that we should proceed with the Man-Made Fibres building and allow Allan Johnson to design the Arts block on the clear understanding that he would be the partner in charge.

By 1952, Lodge had faded from the scene and Allan had his hands full with the four stages of the Engineering block, together with a major extension of the Chemistry and Physics Building, so he appointed an assistant architect, Hugh Ridgeway to deal with day-to-day problems on site. Hugh was keen on climbing, so we decided to have a fortnight's holiday in the French Alps.

Chapter Five

1952–1953: Mountaineering and Marriage

In September 1952, after some limbering up in the Lake District, Hugh and I booked accommodation in the little French village of Samoens. There, we made the acquaintance of Arnold and Eric who were keen on climbing, and suggested that we should join them in tackling Mont Buet, reputed to have the most wonderful views of Mont Blanc. Having no better ideas, we agreed, encouraged by the owner of our little hotel, who made enquiries about a guide, but couldn't find one available. However, he said, a guide wasn't really necessary, because the only difficult rock face had a fixed rope along the dangerous bit. Nevertheless, he recommended that we should take ropes—advice that was to prove invaluable.

We had to wait several days before the weather was good enough, for he stressed the danger of becoming 'benighted' on the mountain. The previous year, even in August, some people had been frozen to death, being unable to get down in daylight. We could believe him, for our very first local climb had taken us above the local lake, which was well below the height of the summit of Mont Buet, and although it was the end of the summer this was well and truly frozen.

On the morning of our departure, the hotelier presented us with the only map of the area he could find, and we caught a local train to a tiny 'Halt' near the base of Mont Buet. There was a respectable cart track leading up to the hut where we were to spend the night. We followed it enthusiastically, until we realised that its meandering route was intended for use by heavily laden horse-drawn carts, with a slight gradient, which was not our idea of mountaineering. So we unpacked our gear, got out the map, and found to our delight that several much shorter footpaths were shown that led ultimately to the hut.

The only problem was to find these footpaths. There was nothing but dense forest and thick undergrowth on both sides and it was impossible to relate the map to the twists and turns of the track without a compass, which none of us had (talk about amateurs). We went on for about an hour before we came upon a footpath that started up the hill in the general direction of the hut with a gradient much steeper than that of the track, which looked distinctly promising.

Our enthusiasm was dampened by Arnold, who declared that it would be safer to continue along the track, and nothing we said would persuade him otherwise. We argued that the map showed that we might reach the hut in half an hour, and Arnold admitted that the track would take much longer, but he was adamant. Because we followed the principle of never leaving anyone on their own on a mountain, Eric agreed to go along with him, despite the fact that this would prolong his exposure to French mosquitoes and flies that had proved to be immune to the ointments we had brought from England. In fact, at that time of the evening, in that thick forest, we were all being plagued by insects and desperate to reach the shelter of any building.

Hugh and I set off, but the path soon died on us. For some time it was very rough going, practically directly uphill, and thick with nettles, brambles, briars, and even fallen trees. We stopped and considered meeting some loss of face by returning to the track, but there was no easy way back. The undergrowth had closed up again and we found that clambering upwards through brambles was marginally easier than climbing down through them on such a steep slope. After another half an hour, the trees thinned out, as did the undergrowth, but we were now faced with a nearly perpendicular ascent and it was getting dark. It was then that we heard a faint shout and glimpsed the torch, which Arnold, having reached the hut, was using to guide us the rest of the way.

When rested and fed, we rather tactlessly complained to the patron about the misleading map. He was not only unsympathetic but called

Chapter 5: 1952–1953 Mountaineering and Marriage

us imbeciles! Didn't we realise that France had been fighting the Boche for six years? All able-bodied men not already in the armed forces had been fighting in the Army of Resistance, and none of them had had any reason to use the footpaths we were looking for. He had only just opened up the hut after it had been abandoned for ten years. Did we really expect maps to be up-to-date? Having pulled our legs, he launched into a lengthy account of his experiences during the war, but with such rapid speech, in what Eric said was a local dialect, we could only catch a glimmer of what must have been a fascinating dissertation. He then stressed that if we intended to reach the summit by noon, as we ought, we should leave his hut no later than four the next morning, and this would allow enough time to reach Chamonix, on the other side of the mountain, by nightfall. The other visitors in the hut, two Germans, decided to leave at 3.00 am to be sure.

After a good breakfast, we set off at 4.00 am into a pitch-black night with just enough light from the stars to make out the footpath. It was more than an hour before dawn began to break, and I shall never forget my thrill on seeing a rosy pink glow on the snowy tops of a mountain range far away to the left. This effect of the sunrise was remarkable, because Mont Buet cast a shadow that gradually sank lower and lower into the valley, allowing more and more of the rosy glow on the snow and ice to appear. As we watched, this became stronger and brighter. We were transfixed, and when we turned to move on, we found that the path over the rocks confronting us was now in such deep shadow that we had to wait until it became lighter. After a while, we could see more clearly, and had only gone a few yards when we came upon fresh blood. Not just a few drops of blood, but a whole pool; more than could be explained by an injured animal. We anxiously debated whether the Germans had had a serious accident, and cast about us to see if we might see them, but finding no trace of anything untoward, we set off again.

The track led us to a rocky plateau with a precipitous slope on one side, and steep contours up which we were climbing on the other. After half an hour, the scene changed dramatically into an arête, where we stopped to consider our options. The severity of our situation gave us something of a shock, especially for Arnold. A few minutes before, we had been safely climbing a very steep slope on a well-defined path. The next minute, this path faded away and we were on the *arête* staring several thousand feet into the valley below. The route (no longer a path) now zig-zagged up this ridge with a precipice on each side until it arrived at a substantial rock face, about a mile away, which rose to the summit. We were all rather sober and got out the map to make sure we were on the right route.

When it was clear that there was no other way up, Arnold announced that he was retreating back to the hut, because whereas we had crampons suitable for rock climbing, he had only hobnails on his boots. He resolutely refused to let any of us return with him, pointing out that it was still only 8.00 am, bright daylight, and downhill all the way on an easy track, so he could not possibly come to any harm. The arête was not quite as bad as it looked, for what had appeared from a distance to be a knife-edge, proved, in practice, to have intermittent stretches of path, although this was so narrow, and the rock so unstable that we did rope up for safety.

Arnold, on the Mont Buet arête, considered his options.

At last we came to the foot of the rock climb and the final stage of our ascent, when all we had to do was to find the fixed rope to get past the dangerous bit. However, winter storms had done their worst, and all that our searching revealed were various lengths of steel cable dangling over open space, well beyond our reach. Thank goodness one or two metal fixings were still in place, so we roped up again and made the climb as

Chapter 5: 1952–1953 Mountaineering and Marriage

best we could but it was a hair-raising experience. We reached the summit at noon, very pleased to be on schedule, and on a perfect day, with blue sky and warm sun. It was glorious. The view of Mont Blanc was fantastic; it was backed by range upon range of mountain peaks capped by snow, with glaciers between them, and most of them almost as high as Mont Blanc itself. We could not see the village of Chamonix because it lay in a valley hidden by the high ground in front of us.

Pausing for breath on approaching the summit of Mont Buet.

We examined the map to decide on the best descent. There were two possible routes: one to the left, and then along the valley to Chamonix, and the other down the valley in front of us, and then over a pass to Chamonix. This second valley opened up invitingly. From the bottom of a scree, not far away, we could see several green pathways wending their way along the valley floor. While climbing with the University Rover Crew in the Lake District we had become familiar with screes consisting of millions of small stones that allowed us to descend at really high speed, running, jumping, and slithering with them, descending hundreds of feet in minutes. We reckoned that we could scramble down the scree in half an hour, and by following the green paths, it should be easy going after that.

How wrong could one be? We were astounded by the time it took even to reach the top of the scree, and although we had realised that it was big we had completely misjudged its scale. It was much further away than we thought, and when we actually reached it, it was more like a 1,000 feet long rather than the 200 feet we had imagined. Instead of small stones, it was made up of rocks varying in size from a cricket ball to a football, which would not allow the sort of free movement we were depending upon. This was just as well, for rocks of that size in motion could have been very dangerous. As it was, we scrambled down slowly and carefully for several hundred feet until we reached the lower slopes, where the angle of repose was less steep and the rocks so large that we literally had to clamber over them or squeeze between them. It took ages, but there was no alternative, because the scree passed between cliff walls, and there was no way round the sides, which in any case were at least half a mile away in each direction.

Beginning the descent into the North Valley

At last, we reached the valley bottom and breathed a sigh of relief as we made our way to those promising green paths on the far side of the scrub. But they were far, far wider than they had appeared from the summit, and instead of firm grass, the green sward turned out to be weed growing in a luxuriant swamp that flowed down the centre of the valley. We had to wade across it to reach dry ground and then follow the stream that gradually emerged from the bog and manoeuvre our way past other quagmires and side tributaries serving what soon became a fast running river; we realised we would eventually have to cross this. The vegetation of coarse long grass on hassocks, scrub bushes and brambles, with not a tree in sight, made it very rough going. The whole valley showed no sign of habitation, so after looking in vain for buildings, or indications of farming, we were astounded and delighted when we actually saw a man! We made our way to where he was clearing the undergrowth, not with a scythe, but with a blade fixed to the end of a very long pole. It reminded me of the weapons used by the peasants when storming the Bastille. Surely, we might now get some information about where we were and obtain some advice about the best route out of this wilderness?

It was now late afternoon; we had been pretty active since 4.00 am and we were getting tired—not to say anxious, but as soon as the man caught sight of us he waved his weapon about and shouted at us to keep away. Why on earth was he being so unfriendly? As we moved nearer to him to attempt a constructive conversation (within the limits of our French), he became increasingly agitated and screamed at us. We thought he must be mad until eventually one word became intelligible—'Vipers!' Of course—whatever work he was doing was disturbing whole nests of them. We then noticed that he was wearing thick leather leggings, and when we glanced down at our bare knees we beat a hasty retreat and abandoned all attempts to talk to him. As we continued on our way, we kept looking for any sign of cultivation or a building that might give some clue to what he was doing, but there was absolutely nothing but endless scrub, and as he proved to be the only living creature we saw in that desolate valley, I remain mystified to this day.

About two miles further downstream, we came across a rough track that came up from the lower valley on our side of the stream, now grown into a torrent at the bottom of a gorge; as we reached it, the track reversed its direction to go over a very respectable bridge. It was now 6.00 pm and we regrouped to consider the next phase of our journey. We were all pretty exhausted and had no more than two hours of daylight left. We could go down the track to the lower valley, but this appeared to be uninhabited and quite desolate as far as the eye could see. Our map didn't cover the area, so we had no idea how far away any civilisation might be. Over the bridge, the track climbed steeply upwards towards the col in the general direction of Chamonix. So, we took a deep breath, gritted our teeth and started uphill. The track was clear, but each of us had to go at his own personal pace—we were all too tired to adjust our steps to match other people's progress. While

Eric led, Hugh lagged behind me, and I wondered if he was going to make it, and what I would do if he could not. While forcing my weary legs to function, I occupied my mind in searching for some primitive shelter we might use if we were benighted, but there was nothing that would serve; not a tree, nor a bush, nor even a large rock as far as I could see.

Looking back after more than fifty years, I am surprised at how much detail I remember with absolute clarity; I suppose it was because I was really scared. Timing was crucial; it was our only measure of distance. I recall that it was twenty minutes after leaving the bridge that I realised that I had not got my camera. I came out in a cold sweat, for it was my most treasured possession—a pre-war single lens reflex Leica that had cost me a small fortune due to the state of supply and demand at that time. I must have left it on the bridge. To go back for it would cost me half an hour, and light was already fading but the run downhill took no more than ten minutes, and it was on the parapet of the bridge where I had left it. Turning back to start the climb again made me realise how little energy I had in reserve; I had to force myself to take each step and when I ultimately caught up with Hugh I found that he was having to take a breather every ten minutes or so. I found that stopping with him made me feel even more tired, so I promised to wait for him further up, and pressed on, and it was just as well that I did.

It was the sheer scale of the country that proved to be far greater than anything we had ever experienced. Lake District mountains were mere hillocks compared with these. Our usual reckoning of 'four hours up and two hours down' just didn't apply here. It had taken six hours just to get down from the summit of Mont Buet, and, here we were, in a state of exhaustion with a mountain range to cross, no real knowledge of the terrain and very little daylight left. As we got higher it became distinctly colder, and when I reached the snowline I became really anxious because the snow had obliterated the path, and all I had to guide me were Eric's footprints that I could hardly see in the fading light. A shout alerted me to Eric's flashlight way above me. He was obviously waiting, so I retraced my steps to find Hugh, who was quite a bit off course, and near the end of his tether—pale, panting, and very cold.

Chapter 5: 1952–1953 Mountaineering and Marriage

We finally got to the top of the col, and with no little relief, saw the outline of a chair lift silhouetted against the night sky. Another hundred yards, and we were gazing across at the faint outline of Mont Blanc on the other side of the valley. While we stood transfixed at this sight, Hugh suddenly started and pointed at his feet. 'Just look down there!' he shouted, and in pitch blackness we saw the twinkling lights of Chamonix over 2,000 feet vertically below us. We forgot our fatigue and stood riveted at the astounding sight. We then saw that the descent would be easy, with a civilized zig-zag footpath and a very essential handrail and wondered what sort of shelter we might find when we finally arrived down in the village in the middle of the night.

It was then that we came upon a miracle:—a Gasthous next to the chairlift that was open—but due to close the very next day! We had never spent such a blissful night.

It was not until the next day, when we arrived back in Samoens and were greeted by Arnold, that we learned more about the pool of blood we had seen. On leaving us the previous morning (it seemed like a week before), he had made splendid progress until he reached the bloody plateau where he was set upon by two large, wild dogs. He was terrified, but luckily there were some loose stones around, and he kept the dogs at bay by swinging his very expensive camera about and throwing stones until he managed to hit one of them on the head with a sizeable rock and they retreated. On reaching the mountain hut, the patron congratulated him on his escape and told him that the Germans had not been so fortunate. The two dogs had attacked them and had bitten one of them in the seat after tearing off most of his trousers. He had been sent to get an injection, as the patron was sure the dogs were rabid. When he bid Arnold farewell, he had his hunting rifle under his arm and was going up the mountain to shoot them. We never found out whether he was successful, or to whom or what the pool of blood had belonged.

During the following year, 1953, Hugh and I decided to tackle the Italian Alps—not that we could claim to have exhausted the French Alps, but the Dolomites sounded exciting and, from the brochures, looked picturesque. We found them to be everything we hoped for. The limestone varied in texture from hard to fragile and we had several nasty moments when handholds and footholds gave way without warning. We learned to climb very carefully and didn't attempt anything risky. The scenery was beautiful and the weather dramatic—the white cliffs were framed by blue skies and billowing cumulus clouds for most of our stay. Occasional thunderstorms made things exciting, because on the mountains, lightning seemed to be awfully close and thunder was deafening. Heavy rain made both the limestone and the short grass dangerously slippery.

After scaling the local peaks, which were not formidable, and finding ourselves in better shape, we made enquiries and learned all about the Marmolada, the highest mountain in Italy, famous for its wonderful views. It was said that on a fine day one could see as far as the Mediterranean. We found two other interested men, including a very nice chap named Ted, hired a car to take us and, in the light of our Mont Buet experience, booked a guide to look after us.

It was a three-hour journey to the base of the mountain, so we had to set off at 5.00 am for a very rough ride over atrocious roads that had suffered from years of wartime traffic and neglect. On the other hand, we were delighted to find a ski-lift working. This deposited us at the base of a huge glacier, where we met our Italian guide, who spoke a little French, but no English. We soon concluded that he must have had a bad time in the war, for he was unnecessarily rude to everyone in the party, which comprised four Englishmen and two charming Frenchmen. He was also greedy; the standard number of people on a rope was six, including the guide, and he had squeezed an extra person into the team. Of course, we didn't realise this at the time, and cheerfully set off as dawn came up, revealing magnificent vistas around us.

At the top of the ice field, the overlying snow was thick enough to hide any crevasses, so the guide started to rope us up. He tied one end to himself, and then proceeded to measure off 15 feet to the next person, tied him up, and then 15 feet to the next until he came to me and Ted. Apart from the guide, Ted was the only one in the team who had an ice axe, so he had been asked to be at the end of the rope. It was now that the guide realised that he had forgotten about the extra member of the party, but he was certainly not going to go to the trouble of unfastening any of the others to give us a fair share. As it took about 5 feet of rope to secure a fully clothed climber, this meant that a mere 6 feet separated Ted and me. At the time, we really didn't mind and treated it as a huge joke, and I was comfortable to be so close to Ted with his ice axe.

The Marmolada icefield was daunting.

The guide took the lead and we managed very well until we came to a huge crevasse, which would have been impassable but for a snow bridge, which the guide was careful to preserve. The far side was steep and very slippery and we were grateful to have crampons on our boots and very glad to be roped together, that is, until we got to the rock face, when our close proximity became a problem. The rock face was vertical, and the only way up was by means of a 'chimney'. This was a fissure large enough for a climber to get into, using both sides for foot and hand holds. The guide took the lead and one by one the others followed, trying to climb at the same rate as everyone else within the limits of the rope available to them—fifteen feet on each side. They seemed to manage without much difficulty, but this was not so for Ted and me, having only 6 feet of rope between us. We found that when I was vertically above Ted there were only a few inches between his hands and the steel crampons on my boots. That meant that my footholds were occupying his handholds, and I couldn't move until he was quite ready.

Our operation became extraordinarily tricky, for if I tried a foothold that didn't work, and put my foot back to where it had been, it was very likely to trap his fingers very painfully. In an effort to avoid this, Ted adopted a system of moving in advance of me so that his hands were just above my feet, and more than once this resulted in his head violently butting my seat and nearly dislodging me altogether. It was just as well that we liked each other, and we started off in high good humour, but after about half an hour of straining, sweating, swearing, and apologising we were getting very tired of it all. When the guide shouted at us, in Italian of course, telling us to 'get a move on', we could have cheerfully throttled him! By that time, we had arranged a system of signals ending in the sequence 'alright then—one, two, three MOVE!' and then we rose in unison eighteen inches or so. It was a very slow climb, with the rest complaining about getting bitterly cold by the time we reached the top of the chimney.

We were able to traverse a huge crevasse, thanks to a snow bridge.

The weather was now closing in, with huge black thunderclouds, well below the summit of the surrounding mountain. The guide called a 'council of war' and we agreed to continue to the top where there was a refuge. However, we did begin to wonder whether we had made the right decision, as snow began to fall, and visibility became so bad that we could only see a few yards in front of us. We were following a ridge, with snow dropping away steeply into what seemed to be a nothingness of cloud and snowflakes. By the time we came to the refuge, we were exposed to the full fury of a blizzard and very grateful to see some shelter—but, apparently, so was everyone else on the mountain. If it had not been for our incredibly rude guide, I doubt if we could have forced our way through the doorway. The tiny hut was crammed with humanity—maybe forty or fifty people all sweating and wet with snow, shouting and singing in French, German, Italian, Spanish, and English, and smelling of garlic and beer. The only thing to do was to eat, rest, and possess ourselves with patience.

There were many who arrived after us who couldn't force their way in, but after an hour or so some people departed, and after two more hours our guide said that we couldn't afford to spend any more time in there. So we went outside to find snow still falling and visibility no better than when we had arrived. In a very bad temper, the guide once more roped up each member of the party in turn until he reached Ted and me. Then he found—would you believe it—that once again, he had forgotten to allow for the extra person. In those conditions, we didn't waste any time arguing, and we let him rope us together on the piece of rope left over.

So, the position was exactly the same as before, except that on the way down the guide stayed in the rear, as the 'anchor-man,' and Ted had to lead the way. In the swirling and drifting snow he had no idea which way to go and had to depend on the unintelligible and very irritable shouts of the guide. I was really scared, for the snow sloped away in every direction, and finding a ridge on a steep bank of snow was very, very difficult.

Somehow, Ted managed it, and arrived at the top of the rock face, with some admiration for the skill of the guide and also some sympathy, because it must have been a strain for him being aware that we couldn't understand a word he shouted. Anyway, we knew we were on the right route when we came upon an Austrian holding on to a rope that disappeared down the gulley that led to the chimney. We waited politely for quite some time until our guide came to see what the trouble was. After a good deal of vituperative language on both sides, it transpired that the Austrian's wife was firmly stuck in the gully and was making little, if any, progress downwards. We waited, and waited, the shouting got louder and more exasperated, and we got steadily wetter and colder until, after what seemed to be an eternity, the guide came to Ted and told him to find another route.

'What, me? In this snowstorm?' protested Ted. In fact, there was no actual conversation as such; it was all a pantomime in sign language, which in retrospect was very funny but didn't seem to be so at the time. So Ted (and I) selected a spot some distance from the gully and gradually edged our way down over the sheer rock face, using our old routine in reverse. This time Ted's fingers were in even greater danger from my groping boots, and we were thoroughly cold and tired, so when Ted announced that he had found a respectable horizontal ledge, we were very glad to edge our way towards the chimney. The whole party followed us until they were all safely assembled on the ledge.

However, when we reached the chimney, we found it occupied by a rope and were disconcerted to see that this was attached to the Austrian's wife, a very large lady who was securely wedged between the two sides of the chimney, just below us. Naturally, Ted and I attempted to help her, but she repulsed us with such a stream of Austrian invective that I was glad I couldn't understand a word of it. We referred the problem to our guide, who was 50 or 60 feet back along the ledge and couldn't hear us. His reply was finally relayed to us. 'Don't bother about the woman—just climb over her!'

More bad weather approaching

So Ted and I proceeded to implement this instruction. As soon as the poor woman realised what we were about to do, she exploded with fury. However, she was not in a position to do as much damage to us as we were likely to inflict unintentionally on her. She must have been horrified when she realised how closely Ted and I were tied together, and I can't imagine what she thought of our 'one, two, three MOVE'

Chapter 5: 1952–1953 Mountaineering and Marriage

routine as we clambered over her body. Of course, we did our best not to put our metal crampons onto any of her more sensitive parts, but because she was filling all the space in the gulley, we were forced outwards into empty space and had no option but to hang on to her for our own safety until we could get back into the gulley below her. It was a good thing she was so securely wedged. The angry protests and abuse continued as each one of our parties clambered over her in turn and reached a crescendo as the last one—our guide—made short work of his passage over her. I don't suppose he bothered very much about where he put his feet!

As we reached the base of the cliff, the snow stopped, the sun came out, and we had no difficulty finding the snow bridge and enjoyed sliding merrily down the ice field to the chairlift.

The following evening, we embarked on the decrepit Italian train (standing room only) to start our fourteen-hour journey through Austria and Belgium. Corrupt railway officials in Austria had made sure we hadn't any foreign currency left to buy food or drink and the ferry across the Channel shared, with British Rail, the job of getting us back to Leeds another twelve hours after that; typical of travel in those post-war days.

So, Hugh and I had plenty of time to sit on the train and wonder what the view from the top of the Marmolada might have been like.

Back in Leeds, one person who had particularly relished the triumph of James' piano recital at Devonshire Hall was the sub warden who detested Commander Evans. Dr MacDougal was a quiet man, very popular with the students, and we got on very well. Having a doctorate from Oxbridge gave him a sense of superiority over the naval officer that he flaunted relentlessly, and James's performance had a delightful subtlety that appealed to him. When he resigned shortly afterwards, the Warden asked me to take his place as sub warden. However, I had no desire to serve under Evans, so I prevaricated, and when the Registrar asked me to become Warden of Woodsley Hall, this gave me a good excuse for declining Evans's offer. I had first met the Warden, Don Appleyard, back in 1946, when candidates were being interviewed for Civil and Mechanical Engineering Lectureships. He had come out of the room with a wide grin and said, 'They've filled the mechanical vacancy—now it's your turn'. So, there seemed to be a curious inevitability in my succeeding him as Warden at Woodsley. I didn't ask him why he was leaving, but when I saw the dreadfully dilapidated state of the Hall, and learned that his House Manager was also leaving, I began to see why.

In 1840, Woodsley Hall had been a stately home built for the Lord Mayor of Leeds, near the top of Clarendon Road, next to the university precinct. It overlooked the city and was considered splendid enough to entertain Queen Victoria when she came to open the new Town Hall in 1858. Huge columns on the front façade were impressive, and the entrance hall, with its gracious staircase, had an air of grandeur. In September 1858, *The Times* said, 'For its size, the house is one of the most complete and richly decorated mansions in this part of the kingdom ... the dining room is one of the most perfect little bijous in the Italian style we have seen. The vestibule and hall are exceedingly well proportioned: the grounds beautiful, and in the most exquisite order'. The Queen was evidently satisfied with her 'bed and breakfast', for she bestowed on Peter Fairbairn a knighthood. But after his death, only three years later, the property deteriorated until eventually it was converted as cheaply as possible into a seminary for the training of clergy. We found the building in an appalling state, and the chapel merely a repository for junk. The gardens so much admired by *The Times* were long gone, although years of neglect and wartime privation had not destroyed the best features of the main building.

The Bursar suggested that, as I was an architect, I should pull it together, and even make plans for its extension, 'for, after all, I had eight weeks of summer vacation still available'. Of course, it was many months before I could prepare sketch plans for its development, and much later still, I learned that the Bursar had quietly shelved these after being advised by someone that they were 'impracticable because they were sited on made ground'—made ground contains material dumped to raise the surface level. This was a relief, for at the time I was embroiled with the problems of the Library ventilation system and the completion of the Parkinson Building (it was some years later, while

Chapter 5: 1952–1953 Mountaineering and Marriage

Mr Chamberlin was working on the main University Development Plan, that his partner, Christoph Bon, built a very attractive extension on exactly the same site as I had proposed without the slightest problem with foundations).

In the meantime, our Assistant Surveyor of the Fabric, Arthur Gilpin tackled the job of repair and decoration with enthusiasm, revealing many of the elegant features of the original house that had been obscured by ramshackle modifications and additions. It was a thrill to see the old house restored to something of its neo-Georgian glory, and I set about dealing with my most immediate problem—to find a House Manager, without whom the Hall could not function. Good ones were very hard to find, but I was incredibly lucky. The first interviewee was a Miss Eleanor Margery Fawcett, a very attractive young lady with impeccable qualifications; a Cordon Bleu from Paris, experience in hotel kitchens, and Assistant Catering Officer at Manchester University, where she had recently slipped in the kitchen and was now recovering from a broken wrist. I couldn't believe it—she was too good to be true, and obviously qualified to have a far more important job than I was offering. So, I took her to Woodsley and we made our way through the clutter of builder's materials down to the tiny, ill-equipped kitchen, where the kitchen staff had left the obsolete stoves, ovens, and a miscellany of pots and pans in careless abandon. It looked dreadful; I saw her face 'drop' and it was clear that she had no intention of taking the job, for her first question was, 'How do I get back to the station?'

I explained that there were more than two hours before the next train to London, so there was time to show her the Manageress's flat, which comprised a magnificently ornate bedroom with a plaque recording that it had been slept in by Queen Victoria during her state visit to Leeds, together with a small sitting room and a private en-suite bathroom. She appeared quite unimpressed by this last ditch attempt to entice her, but agreed to return with me to my office in the Parkinson Building, where we had a cup of tea, and I put forward every argument I could muster, until she consented to 'consider the position further'. Having had no other response to the advertisement, my relief can be imagined when she telephoned to say that she would give it a trial, and astounded everybody by arriving the very next day and promptly taking responsibility for the reconstruction of the kitchen quarters and all the associated service rooms. This positive attitude evidently sprang from her experience during her War Service, when she had been a Red Cross Commandant in charge of a Military Reception Centre in East Anglia, caring for wounded soldiers and airmen, with a certificate signed by King George and Queen Elizabeth to prove it. The Centre was on the coastline, and on one occasion, a German fighter machine-gunned her while she was walking on the front—luckily for us, he missed.

It was perhaps the most important issue of my career, for it changed my life completely. Margery was a treasure; the domestic staff loved her, for they found that she was quite prepared to take over their work when anyone was sick, and without reservation. She insisted that I should check the housekeeping accounts every week, although her arithmetic was more reliable than mine. After the first year, she was the only House Manager in the university to keep within her budget, although theoretically, Woodsley was too small to be financially viable. The University Auditor was very impressed and told me, confidentially, that at one of the other halls the manager would ring the butcher and say, 'send me enough meat for 150 students!'

Margery is the only female present in this shot of Woodsley Hall.

Chapter 5: 1952–1953 Mountaineering and Marriage

However, her remarkable control of costs was not at the expense of the catering. The students compared the high quality of her meals very favourably with other halls, and most of the comments in the complaints book that lay on the entrance hall table were frivolous. Even so, the House Secretary, Peter Sykes, would never let Margery near it. Like all members of the Hall he was determined that nothing should upset her, so he took it upon himself to convey any 'complaints' to her tactfully (in this way, it was agreed that their treasured bottles of tomato ketchup should be on the dining tables only at selected times and with certain menus).

By the end of the first term, I had plucked up enough courage to invite her to the Engineer's Ball and to introduce her to Prof Evans and all my colleagues. I shall never forget my dear friend Dr Brittain whispering in my ear, 'My God, Geoff, you can certainly pick 'em!' We got home about two o'clock in the morning, exhausted. The next day I crept sleepily downstairs to find that she had been up since six o'clock getting breakfast for the men; the cook hadn't turned up. She was as bright and fresh as if she had had a full night's sleep.

It also turned out that my appointment of John Woodhead as sub warden was inspired. He had lived with me at Devonshire Hall and was very glad to get away from Commander Evans. A lecturer in Phonetics, and a graduate of Oxbridge, he seemed to 'know all the answers' and succeeded, with the Student President and the House Secretary, in settling most problems without troubling me. It was just as well that he did, for my days were so filled with giving lectures, supervising drawing office work, meetings and appointments with departments and our consultants that I always came 'home' very tired indeed and was grateful for the warmth and friendliness I found there.

We were also well served by our Student President, Roger Kirk who was worthy of the title setting the tone of the Hall with quiet authority and good sense. With Peter Sykes, and a tiny student committee, he ran all social events without recourse to me except for approval and advice. It was Roger and Peter who interviewed freshmen who applied to join the Hall, dealt with disciplinary matters, and took care of our guests at the weekly formal dinners. These meals were quite different from those at Devonshire Hall, at which everyone was required to wear a black academic gown at evening dinners and Sunday lunch. At Woodsley, we never wore gowns, which I regarded as an unnecessary expense; the only formalities being a collar and tie and good manners. Coffee in the lounge was followed by informal discussion with the guests in a relaxed atmosphere.

In this way, we entertained Stanley Ellis who fascinated us with his knowledge of English dialect. He expertly helped the police to track criminals by their voice characteristics and language and became internationally famous for his part in the Yorkshire Ripper case, during which murders continued to be committed because the police failed to act on his advice. A man with a Wearside accent claimed to be the Ripper, and although Stanley warned them that this was a hoax, they continued to waste time and resources searching in the wrong places. In the meantime, Sutcliffe, who lived locally and had been interviewed and cleared nine times by the police, continued to kill prostitutes until he was finally apprehended and admitted to thirteen murders.

Another memorable guest was my friend Dr Robert Brittain, a senior lecturer in Forensic Medicine with degrees in medicine, law, and science. His term as a Research Fellow at Harvard resulted in a paper on the use of tattoos in identification, to be followed by 'singular monumental ground-breaking work' on the careers of sadistic murderers. This provided the template of subsequent investigations and became known as 'Brittain's Sadistic Murder Syndrome'. Robert delighted us with somewhat gruesome experiences, such as the time when he was asked to check on a suspicious smell emanating from an ancient cottage. He found a frail old woman sitting up in bed with her husband lying alongside her. When Robert exclaimed that he must have been dead for at least two days, all she said was, 'Well, I nivver, but I thowt 'e looked a bit queer!' And on another day, when doing an autopsy, he was told that the woman had fallen downstairs while carrying some scissors. They had not done any serious harm at the time, but they closed as she pulled them out and fatally severed a major artery.

Robert finished his reminiscing on a lighter note, about one of his trips to Scotland, where his invalid mother lived. He visited her every other weekend in his prized possession—a pre-war Jaguar. He always used the A1 highway, founded on the old Roman road, which ran due

Chapter 5: 1952–1953 Mountaineering and Marriage

north from London up to Edinburgh. It had never been widened, and only took one line of traffic in each direction. Moreover, during six years of war there had been very little maintenance, and army vehicles had played havoc with the surface, having to carry all the heavy goods traffic to and from Scotland. During the daytime, traffic was continuous, and frequently slowed down to a crawl because it was impossible to pass large trucks on such a narrow road. So Robert travelled through the night. The 200 miles along a straight road were very boring, for in those days there were no car radios to ease the monotony, so one day he persuaded his friend, John Ireland, to keep him company. On their way home on the Sunday night, a car came towards them with headlights full on. At first, this didn't matter, for it was a long way off, but as it drew closer, and didn't dip its lights, as was required by the law and common decency, Robert flashed his lights to alert the other driver. As this had no effect, he pretended to be dazzled by the headlights and drove straight at him, swerving away at the last minute.

A little time later, they found themselves being followed by a police car with a siren and were forced to pull up. Robert was accosted by a very angry policeman who demanded, 'Are you the maniac who nearly drove into me?' To this, Robert retorted, 'Are you the idiot with headlights that nearly blinded me?' After this the exchange went:

'Like that, is it? What's your name?'
'Brittain'.
'And where are you coming from?'
'Scotland'.
'Where are you going?'
'England'.
Policeman, turning to John,
'And what's YOUR name?'
'Ireland'.

At this, the policeman's anger exploded, until documents were produced that showed that Robert was being perfectly honest and not trying to ridicule him. He cooled even more significantly when Robert commented that his detective friends in the West Riding Police who worked with him every day would not be impressed to hear about the officer's behaviour in this incident. They parted amicably enough.

It was a tradition among the men's halls that raids could be perpetrated on each other's premises with the object of securing a trophy of some sort. The Woodsley team brought back a ship's bell from Devonshire Hall, which did not please me, for I knew it was from one of Commander Evans's ships, used to announce dinner, and so I was obliged to invite him to a formal dinner to make an apology. Inevitably, there was a responding raid from the Devonians, but we were ready for them and they were repulsed without casualties, for the struggling was good-natured and regarded as fun. Nevertheless, one of the raiders suffered a superficial injury, and I was persuaded to get my car out and take him to the Infirmary for some stitches were needed. The doctor on duty in the A&E department was an old member of Woodsley Hall and delighted us by using more iodine on the wound than was strictly necessary. After howls of protest from the patient, he apologised for his clumsiness.

My social life continued apace, with attendance at university functions, formal balls, public lectures, and Scottish dances filling my diary. But, I now had someone who seemed willing to partner me at these events and made me thankful that I had got round to learning ballroom dancing. Margery made a partner of whom anyone would be proud, and we attended several important events, such as the Union Ball during which she was presented to the Duchess of Kent. Then, on the 14th February 1954, I kissed her (in accordance with a resolve approved by my father in the 1930s that I should never kiss a girl unless I intended to marry her). A month later she agreed to marry me.

My weekends with the Rover Crew were replaced by outings with Margery, except one Sunday when the washer-up failed to arrive to deal with the dishes from lunch. It took Margery and me all afternoon to clear the kitchen of all the dirty pots, pans, cutlery and crockery left from a four-course meal for fifty students. I resolved that by hook or crook the hall was going to have a washing-up machine. At that time, the only types available were huge commercial machines for large kitchens, or tiny domestic ones that were useless for fifty people, but I realised that my prestige was at stake, and by summoning up all the resources available to the Planning Office, I eventually did get hold of one that served us well. By 1954, technology had spread to the Warden's office by the installation of that thing of wonder, a television set.

Chapter 5: 1952–1953 Mountaineering and Marriage

The BBC transmission, the only one available, operated for a few hours each day with a black and white picture of very poor quality (405 lines in VHF before it was improved to 625 lines on UHF). Breakdowns were frequent, but the 'interludes' that filled the gaps with scenes of running water or seascapes, were accompanied by delightful music and were a pleasure to watch. Students were invited, but only for special programmes selected by the President.

We decided to get married in August. Margery's family lived in the south of England and had arranged an impressive wedding in London with a reception at the Hyde Park Hotel. Her sister Audrey was Maid of Honour, her cousin Gillian and my cousin Jackie were bridesmaids, and John Woodhead was my best man. After that, our perfect wedding day threatened to become a disaster when we arrived in Zurich and found that the hotel had cancelled my reservation for our wedding night because I had not specified that we would be arriving after six o'clock in the evening! I shall never forget my frustration and embarrassment as I compared my miserable performance with Margery's elaborate and successful arrangements for our wedding—I couldn't even book a bedroom successfully. A large conference had taken over every room in the city, but after an interminable wait, a room was found for us in another hotel, and when we arrived in Zermatt the following day, things went well.

After a warm homecoming to Woodsley, I discovered that one of the prerequisites of marrying the House Manager was that I could now share her private bathroom. This was a luxury, for hitherto the only bathtub available to me was one in the middle of a washroom for students' use, with no privacy whatsoever. There were no showers anywhere in the building. This idyllic state soon changed when Margery announced that she was pregnant. This meant that we would have to leave Woodsley, for our rooms were the only accommodation that could be offered to any house manager taking her place. John took over as warden, and Major Currer Briggs offered us a flat in his home as a temporary measure. The terms of the lease excluded dogs and children, but the Major turned a blind eye to Margery's cocker spaniel, although he made it clear that we should have to move out before the baby arrived.

So, the provision of a home proved to be the first major challenge I had to face in our married life; I had to promise faithfully to find a house before Margery produced the baby, without having the slightest conception of the tribulation Margery would suffer on her part, nor what stress my promise would entail.

The search for a home was a formidable challenge, for very little had been built since 1939. Licences and building permits were required for building in the cities, and many people were looking for homes in the countryside. This suited Margery, who didn't fancy living in Leeds, but as country cottages were snapped up before they had been advertised, buying one wasn't going to be easy.

Margery was not feeling well, and I had to go it alone, touring public houses in the hope of gleaning information about anything likely to become available. The locals regarded me as a foreigner, for my Yorkshire accent was distorted after five years in London during the War; I ordered sherry instead of beer and had neither the confidence nor the money to stand a round of drinks, so my endeavours were a pathetic failure. I drew a complete blank until I bought some eggs at a farm near Kirkby Overblow, at which I discussed my predicament with the sympathetic farmer's wife. She thought that two elderly ladies in the nearby hamlet of Kearby *might* be thinking of moving, but she emphasised that I must be very tactful if I approached them, for it was only hearsay.

No time was lost in fetching Margery for a reconnoitre. From the roadside, we could see that their ancient stone cottage had lovely gardens and a wonderful view down the Wharfe Valley. It was idyllic, and we wondered whether there was any real hope of securing it. When the two elderly ladies came to their door it was clear that they wondered too, astounded to hear that perfect strangers had discovered their private intentions. They looked at us suspiciously until we explained how we had learned quite innocently about Pasture Cottage. They already had a list of friends and acquaintances to whom they had promised first refusal should they ever sell. They hadn't *decided* to do so but agreed to put our name at the bottom of the list. It seemed a forlorn hope, so with only six months left before our deadline, the search had to go on.

Meanwhile Margery's doctor discovered that her ill health was due to a serious case of anaemia and insisted that she must have a holi-

Chapter 5: 1952–1953 Mountaineering and Marriage

day in the sun, without delay. That meant travelling south, for it was early March. When cousin Cecil heard of the problem, he spoke to a travel agent friend of his, booked us a holiday in Majorca, made comprehensive travel arrangements and sent us airplane tickets, before we had even begun to think about packing. Commercial air travel was still very primitive in 1955. The departure lounge at Heathrow was made from semi-circular sheets of corrugated iron, although work had started on a permanent terminal building. All the runways were of grass, but recently the Queen had ceremoniously laid the first slab for a concrete runway.

Our aircraft was a dilapidated Douglas DC3, the 'workhorse' of the RAF during the War, having carried hundreds of soldiers on benches before being fitted with passenger seats. The whole plane shook as the ancient radial engines struggled to get it off the ground; it was far more exciting than any flight today. As we approached Palma from the sea, the plane banked so steeply and flew so low that I thought the wingtip would touch the water before it levelled out and landed with a jolt that must have strained the feeble undercarriage. As we lurched across the rough field towards a wooden hut, I wondered if the pilot had made a mistake, until a windsock showed that indeed, this was the purported airfield.

In the corner of the field, we were warmly greeted by the robust owner of our hotel, who asked to be called 'Gerty'. She drove in pitch darkness over rough country roads to Soller, on the far side of the island. Gerty prided herself on spoiling her guests with good food and drink, so after two weeks, Margery was restored enough to visit the house where Georges Sand and Chopin had lived. In the meantime, we painted in brilliant sunshine. My experience was limited to tinting architectural drawings, so my efforts were fastidious and weak. Margery used strong colours, with firm brush strokes, and achieved far better paintings than I, but she was never satisfied, and annoyed me by always tearing up her work shortly after finishing it. Until that holiday, I hadn't had time to read fiction; hitherto all my vacations had been spent camping, climbing, or studying for exams. Margery just couldn't believe that at the age of thirty-four I had never heard of Agatha Christie, nor read any detective stories. So, when we discovered a little English lending library on

Majorca, my holiday acquired an unexpected richness; it was the first 'restful holiday' I had ever experienced and it was a very nice change.

Our hostess occasionally mentioned 'her time in Holloway', but it wasn't until years later that we discovered that she had not been referring to the residential district in London, but to the women's Holloway prison. She had resolved to give her daughter an education at an expensive private school, and to raise cash she had joined a smuggling gang, which in those days was a very lucrative business. Cameras, watches and jewellery were practically impossible to get except on the 'Black Market'. Since Britain had spent all its foreign reserves in fighting the war, tight restrictions were imposed on anyone taking money abroad, and a tax of 60 percent was slapped on all luxury items. So, Gerty's enterprise was particularly profitable. She put secret panels in her Rolls Royce and made regular journeys to Switzerland 'for the benefit of her health', returning with loads of Swiss watches. She sometimes took her daughter, always varied her route, and on one occasion invented an illness so serious that she had to get the AA to bring the car back to England for her, together with the loot. As a precaution, she made regular telephone calls to the gang en-route, so that when, inevitably, she was caught and failed to make the prearranged call, the gang were able to scatter. Her resulting prison record made further smuggling in England out of the question, so she moved her operations to the Mediterranean, where France, Italy, and Greece were desperately short of cigarettes, silk stockings, and nylon underwear. In her fast motorboat, smuggling was easy because these war-torn countries had no money to spare for Customs patrol boats. She was making a very comfortable income until intercepted off the coast of Italy by pirates in a German E-Boat armed with heavy machine guns. The loss of the whole cargo of contraband was serious enough, but the pirates scared her so much that she retired, and now she was living a quiet life as a hotel keeper, in which she was obviously going to be equally successful.

We didn't expect any drama on the flight home, but the weather over France was stormy, and because the plane was not pressurised it had to fly through the clouds rather than above them. Sudden downward plunges and bumps made Margery worry about a miscarriage, and she was beginning to feel ill even before the pilot began to descend, during which announcements about an emergency landing didn't help. Thank goodness we were only making a pit stop in Paris, because headwinds had made us short of fuel, and when we took off again, the thunderstorms had cleared.

In my renewed search for a house, I completely abandoned 'pub-crawling', but my alternative technique resulted in the purchase of a ridiculous number of eggs without any result. At last, the owners of Pasture Cottage decided to sell and were negotiating with maddening slowness down the list of prospective buyers. Many sleepless nights later we heard that we were the only ones prepared to meet the asking price, and the sale was agreed, on condition that we should have possession by the end of September—a very tight schedule. But, the two ladies kept their word, and it was just as well—for when we got the keys on the 27th September and saw the empty house for the first time as property owners, the excitement was too much for Margery in her delicate state. We rushed straight to the Maternity Hospital where Susan was born the next day, and as the removal men arrived with our furniture that afternoon, I found that I had kept my promise with about one hour to spare.

Chapter Six

1955–1962: Pasture Cottage

The pregnancy should have been straightforward, but it certainly was not. Margery had had the prescience to consult a Harley Street surgeon, who had warned her that it would be dangerous to attempt a normal birth and suggested that she would be well served by the head of the Department of Obstetrics at Leeds University. However, we found that very eminent physician, Professor Claye, was approaching retirement, and was far more interested in being eminent than in actually practising what he taught. He promptly referred Margery to his senior member of staff, Miss Lister.

When Margery told her of the advice she had been given in London, Miss Lister made it very plain that no-one in Harley Street was going to tell *her* what to do; *she* would be the one to decide, when the time came, whether a caesarean operation was to be performed or not. So Margery approached the birth with far more than the usual trepidation and it was not until the last minute that it was announced that a caesarean was the only safe procedure. Of course, the baby was perfect, but Miss Lister hadn't finished. For some inexplicable theoretical reason, she kept the baby away from Margery as much as possible, until she became agitated and demanded to see her offspring. Even then, Miss Lister brought in the baby, hanging by its heels, saying, 'See—she's quite tough!' It was cruel and unnecessary, and we never forgave her.

I resisted the temptation to attempt some retribution, but it so happened that some years later Miss Lister asked me to provide accommodation for an electron microscope. These exotic instruments were at the cutting edge of scientific research, and as we had just installed one for Irene Manton, Professor of Botany, Miss Lister thought she would like one too. The installation had been very expensive, requiring a dust-proof laboratory with full air-conditioning, difficult to contrive in the old Botany house with wooden floors and leaking windows. However, Miss Manton was a Fellow of the Royal Society—a rare distinction of which the university was very proud. She was a brilliant scholar of international repute, so I pulled out all the stops to give her what she needed. Her fame was such that American scientists invited her over to New York University and offered her the use of their electron microscope while hers was being serviced.

I really couldn't do anything for her friend, Miss Lister. Funds for such expensive equipment came from Research Councils, and Miss Lister hadn't even got a promise of a microscope yet. She was comparatively unknown, and competition was keen. When, at long last, a microscope was promised, it was only on condition 'that a suitable laboratory could be provided.' There was the rub; this involved the considerable cost of converting a cellar in her old house into a Grade A laboratory. Compared with demands from other departments, the committee considered her priority low, and she had to wait a long time before the work could be included in my programme. By that time, the university had the distinction of having seven electron microscopes—more than any other university in the country—so she finally got one and, would you believe it, very shortly after it had been installed, she abandoned it to take up a post in South Africa.

Professor Manton was a colourful character, worthy of special mention. She had been in the habit of sending complaining letters every few days to my predecessor, Mr Ilett, who simply ignored their forceful language, and put them in a file called 'Botany complaints'. I then realised that she had plenty to complain about. All her accommodation was in a very old 19th century terrace house, quite unsuited for research in a scientific subject. I hadn't met her, but I had seen her, at a distance—indeed, one could hardly miss her. She was a formidable figure as she strode about the place with a perpetual frown on her worried face as though the whole world was against her. Other members of the administration seemed to be afraid of her—presumably, they too had received angry letters of complaint. As the Botany Department was housed immediately opposite my office, and I always parked my car outside its entrance, there was little chance of avoiding the good lady, and I thought we had better meet.

She brusquely showed me over her laboratories and emphasised her complaints about inadequate heating, lighting, and electrical services and the problem of keeping things clean. Some matters could have been dealt with if Ilett had simply lifted up a telephone, and I could sympathise with her irritation; others needed considerable work and expenditure. When she found that, at last, someone was taking serious notice of her problems and was prepared to help, she was so pleased that her whole demeanour changed; her demands not only became reasonable but were also reduced to a manageable list that I could cope with. So, we parted on terms of mutual respect, and I was glad to close the 'Botany complaints file' forever. Although Professor Manton made it clear that she did not 'suffer fools gladly', I suppose that, to someone with her brilliant intellect, there must have seemed to be an awful lot of fools about. Some people called her a 'harridan', but as the years passed I learned something of her true character, what a kind and generous person she was, and how much her students loved her. Her valuable collection of beautiful modern paintings was left to the university.

The following year, 1956, saw the start of the last of Lodge's designs—the Man-Made Fibres Building. This had been sponsored by the Clothworker's Company, and financed by Courtaulds, a firm pre-eminent in the textile industry at that time. They had persuaded the Duke of Edinburgh to lay the foundation stone. As His Royal Highness had a reputation for amusing himself by searching for and exposing of any 'humbug' on his official visits, we made sure that everything was well organised, and we were very carefully coached in the etiquette required for the occasion that was to be presided over by the Princess Royal.

All went well until the master mason spread the mortar onto its bed and invited the Duke to lower the foundation stone that was suspended from the scaffolding by block and tackle. The Duke pulled on the chain, but the stone went up instead of down. There was a horrible silence, and some nervous titters as the poor mason was firmly pushed aside by the Site Foreman who spent what seemed to be an eternity

HRH Duke of Edinburgh laying the gravity-defying foundation stone of the Man-Made Fibres Building.

fumbling with the tackle. At last, he handed over the chain for the Duke to pull, but the stone rose even higher! For a few tense moments there was an embarrassed silence, while we imagined the worst, but when the Duke burst out laughing, everyone joined in with relief. It must have been boring for him to come to a provincial university merely to lay a foundation stone, and now he could relish the thought of relating this ludicrous episode when he got home again to the Royal Family.

Professor Speakman was determined to continue the long tradition of patronising the fine arts by having a sculpture somewhere on the façade of the building, and Lodge insisted that, as architect, he should be allowed to design it. When he asked what the motif might be, the professor suggested, 'man making, with great difficulty, what nature achieves so easily'. Lodge then submitted a series of drawings showing an old man using his stick to strike a rock from which a liquid flowed—the symbolism of which neither the professor nor I could understand. Each subsequent proposal was quickly rejected and after frustration at the lack of progress, Professor Speakman gave the commission to Mitzi Cunliffe. She was an American sculptor of international repute who

Chapter 6: 1955–1962 Pasture Cottagee

Sculpture for the Man-Made Fibres building by Mitzi Cunliffe.

produced a splendid piece depicting a pair of hands holding a cat's cradle of fine filament, and this now crowns the front façade.

Until that date, our most notable and valuable sculpture was a war memorial by Eric Gill, commissioned just after the First World War. It showed Jesus driving the moneylenders from the temple, and was at the foot of the South façade of the Great Hall, fronting on to University Road. It had immediately created controversy, I suppose because the figures wore contemporary coats and hats rather than traditional biblical clothing (Ilett told me that the Registrar, Mr Wheeler planted a creeper to obscure it, and the Vice-Chancellor, Sir Michael Sadler, came down on a Sunday morning and cut it away with a large pair of scissors). The sculpture was made of limestone, and when we found the effect of acid rain becoming evident, we gave it the necessary cleaning and moved it to the foyer of the new Arts Block, which housed the Department of Fine Art.

In the planning of this building, Allan Johnson did not let me down; it was generally regarded as a very satisfactory design. This was indirectly due to Prof Evans, who had insisted that his new Civil Engineering department should have a 'focal point', where students would foregather naturally, next to the professorial suite, with a coffee bar, notice boards, seating, and so on. This arrangement was a major break from the usual pattern of dreary long corridors and empty hallways, and was so significantly successful that I had no hesitation about specifying similar focal points in all future departmental buildings. The Arts Block housed a series of relatively small departments, such as Russian, Spanish, Phonetics, and others, and each had staff rooms grouped round a common room, which gave a friendly intimate atmosphere, despite being within a building of monumental scale.

It was most unfortunate that the piecemeal allocation of grants in the 1950s resulted in our having to build the Arts Block in four separate phases, with unbelievable complications and extra expense. When completed in the 1960s, the Arts Block was to make an important contribution to the Vice Chancellor's resolve to correct the imbalance inherited from the original Yorkshire College founded in the 1880s, in which large departments of science and technology still dominated the much smaller faculties of Arts and Social Sciences. Sir Charles recruited several illustrious names, including Quentin Bell, nephew of Virginia Woolf, and Sir Laurence Gowing as professors of fine art. He also persuaded Eric Gregory to endow a series of Fellowships that would allow well-established artists to live and work in the university on condition that they would mix in with students and encourage their artistic aspirations.

The first Gregory Fellow appointed was an architect, Reg Butler, who had distinguished himself in 1953 by winning the prestigious international competition to commemorate 'an unknown political prisoner'. His design was for a gigantic metal sculpture to be erected on a prominent site somewhere on the continent, but the work was never commissioned. The vice chancellor was keen to make this first Fellowship significant and asked me to find a suitable flat and studio as soon as possible. I saw no problem with this—we could make space for a workshop in the men's locker room, in a busy central area near the Great Hall that was always crowded with students, and we could provide a flat just across the road in Beech Grove Terrace. To everyone's surprise this did not suit Reg at all; he insisted that 'the flat must be immediately next to the studio so that I can call to my wife to bring me coffee or hold a piece of metal while I work on it'. I scoured the campus for a solution, but there was no way we could meet this specification, so I was driven to look at our residential sites and suggested that a flat could be made in one of the annexes to Devonshire Hall, with a workshop in an adjacent outbuilding. Reg didn't bother to look at my proposals because he said he did his best work in the middle of the night, and he thought that the screech of the electric grinder and hammering on the anvil might disturb the students (and what about the neighbours, I wondered?)

Every suggestion I made was rejected, and he finally proposed that, as an architect, he might design a flat and workshop himself on 'some piece of land where I would not be a nuisance'. I wondered how long that would take! By this time, a large chunk of the Fellowship had elapsed, and the VC was being criticised in Senate for paying a salary to a Fellow who had not yet shown his face in the university. So, when I heard that Reg had strained his back and had agreed to spend the rest of his tenure in a room in the middle of the campus, making little models in clay, I evinced more relief than sympathy. You can imagine my relief when other Fellows, such as Trevor Bell, Austin Wright, and Alan Davie turned out to be poets, requiring nothing more from me than a room, a desk and a chair! Sometime later, we created an artist's studio/flat that was occupied in succession by several Fellows in painting, including Terry Frost, who became famous and was awarded a knighthood.

The Vice Chancellor drove me hard, but I thought the world of him, especially when he supported my efforts to develop my course in Architectural Engineering. He tried to persuade the City College to transfer their School of Architecture to the university, but as this school was 'the jewel in their crown', his attempts failed. Eventually, he found a benefactor, who endowed a Chair of Architecture, to be dedicated to Hoffman Wood, a prominent Leeds architect who had founded a trust to provide scholarships for architectural students. The incumbent was required to give six public lectures and take some interest in my course on architectural engineering.

The first to hold the chair was Sir Basil Spence, who had received universal acclamation for his newly completed Coventry Cathedral. His public lectures describing that building were very popular, but he had absolutely no interest in my course, making only one perfunctory visit to the department, and that was without my knowledge. The following year, Sir William Holford took over the chair, earning my respect in every way, delighting us with his reminiscences. He was evidently quite brilliant, for he was only twenty-six when he was appointed Professor of Architecture and Civic Design at Liverpool—one of the oldest and most prestigious schools in Britain. He told me how, while sitting at his desk only a few days after taking up his post, he took a telephone call from someone who evidently felt himself to be very important. A local school had decided that the Professor of Architecture should present the prizes at its forthcoming speech day.

William suspected that he might not be the person they had in mind, for this was an honour usually accorded to people who had achieved public eminence, and he tried to wriggle out of it, but when the pompous caller persisted, and forced him to admit that he was, in fact, the Professor of Architecture, and refused to take no for an answer, he reluctantly gave way.

The school was not in Liverpool, and William travelled by train on the appointed day to find a group of dignitaries, including the Mayor and Lady Mayoress, assembled on the station platform together with a crowd of press photographers. Evidently, they were there to give a civic reception to some distinguished visitor, so he stood aside to see who this could be. As the platform emptied, it dawned on him that everybody must have been waiting for the guest speaker so he went up and modestly introduced himself. When they saw how very young he was, their faces reflected horror, disbelief, and embarrassment. He was bundled unceremoniously into a car, taken to lunch (where he was fascinated by the Lady Mayoress's gold chain, dangling from her ample bosom, repeatedly dipping into her soup), and then to the speech day ceremony. He had prepared an appropriate and witty speech, of which he was quite proud, but those who preceded him evidently no longer regarded him as chief speaker for they droned on until the audience became fidgety and restless. This exacerbated the way in which he had been treated at lunch, where he had been ignored completely, so when he rose at last he told the assembly that in view of their forbearance during a long ceremony, his speech would be very short indeed. He then announced, with tongue in cheek, that the Governors had agreed to grant a full day's holiday this year, instead of the usual half day. With resounding cheers from the boys, and a stunned silence from the shocked occupants of the platform, young William departed with dignity and was never asked there again.

Sir William did me the courtesy of visiting and approving my course on architectural engineering.

Baby Sue and her mother arrived at Pasture Cottage with as much ceremony as I could muster, and when I enthusiastically flew a union flag on our flagpole (of all things to find in a country cottage garden), the locals must have thought their new neighbours, who they had not yet met, must be 'a bit peculiar'.

Pasture Cottage overlooked the river Wharfe and Wharfedale

Kearby was a Scandinavian settlement of around 1,000 years old; a hamlet consisting of one large farm, two rows of worker's cottages, and some dozen houses scattered round about. It was on the crest of a ridge overlooking the Wharfe river valley. Pasture Cottage had been built into the side of the hill and had magnificent views up the valley. It had grown by the accretion of bits and pieces, which, by the dictation of the steep slope, were each on a different level, requiring steps everywhere within the house. An old and beautifully lined well, complete with winding gear and bucket, had served the cottage with fresh water and now provided an attractive water garden, complete with a waterfall.

Our predecessors had been retired, and all their energies and resources had been concentrated on that garden. Their enthusiasm had been boundless, and sometimes misplaced, making even simplest maintenance a challenge. For instance, the orchard and vegetable garden were surrounded by a boundary comprising a substantial privet hedge, backed by a laurel hedge and then a row of poplar trees that had grown into each other so intricately that balancing on a stepladder to trim all three of them together was an annual nightmare for me.

Margery was an enthusiastic gardener, but she had a new baby, two dogs and a husband to care for. So, maintenance started with the closure of some flowerbeds. Even so, there remained lots of flowers to pick, and the workroom next to the conservatory was well used. We also grew lots of tomatoes in the large greenhouse next to the garages, so we were kept pretty busy. We couldn't have coped without the occasional help and advice of John, who lived a little way down the hill. He was one of a 'dying breed'—a genuine countryman, who was employed by the local council to keep all the hedgerows trimmed and ditches cleared for miles around. He knew all the local gossip; where hedge sparrows had their nests, and which ones had been taken over by cuckoos. Thorn hedges were the harbingers of spring, the first to show fresh green shoots, and John was bitterly critical of farmers who sought to gain a few extra square yards of land by cutting them down and planting ugly barbed wire fences instead. We agreed with him.

Eventually, Margery began to realise that the remorseless fatigue she was suffering was abnormal, but when Dr Mitchell could not find anything wrong, he tried to reassure her by saying, 'All housewives complain of being tired'. He even suggested she might get a job. Only when she pointed out that her neck was swollen did he realise that she had a large in-growing thyroid gland. The lengthy operation to cut the goitre down to size would not have been possible before the war, when drugs to keep her unconscious for a long time were not available The surgeon expressed his satisfaction with the operation, but it took so long for her to recover that he finally admitted that he had taken away too much of the gland and prescribed thyroid supplement for the rest of her life. In the interim, we managed to get help from au-pair girls from Germany, and Gudrun, who was an angel, is still a staunch friend of the family.

Although Kearby was such a tiny hamlet, it was surprisingly full of unexpected characters. An old woman who lived opposite us had never seen the sea, and in her very long life had not been further afield than Wetherby, only four miles away. Next to her lived Brian Youatt, chairman of the Yorkshire Electricity Board, whose wife Quita, who was from New Zealand, became a firm friend. She was a successful breeder of British Bull Terriers, going as far as America to judge at dog shows. Her brusque personality reflected the characteristics of the dogs she bred, for she was a woman who always expressed herself strongly. When the mineworker's strike forced Brian to devise a system of rationing electricity, with power outages for everybody, including his own home, her language was unprintable.

The first neighbour we met socially was Helen Crockatt, who lived next door; she had called round while our furniture was being moved in, with a cup of tea and offers of help. As I welcomed her into the cottage we stared at each other in astonishment, for we had met once before—in Majorca of all places, while we were staying at Gerty's hotel. Helen had a daughter Jan, who was a year older than Sue, and they became friends when they both attended the preparatory school in Harewood village.

It was Helen who told us all about Gerty's history, and insisted that we should meet another of her friends, 'La Contessa', who lived unobtrusively in a tiny cottage opposite our house. The reclusive behaviour of this lady was due to the fact that she was mistress of an eminent barrister who would have been debarred if their secret affair had become public, but she did invite us to tea, and as Margery couldn't go at the last minute, I was entertained by this elegant beauty all to myself. She declined our invitation to a meal and we never glimpsed her again in all our seven years at Kearby (many years later, we learned that the barrister had obtained a divorce, married the countess, and had followed Gerty's example by opening a hotel on the Mediterranean coast).

The next surprise was to find that our predecessors at Pasture Cottage had had professional advice on gardening from Leeds University, in the form of a jolly lecturer in horticulture, Miss Turner, who lived nearby in Kirkby Overblow, with her bosom companion, Miss Mitchell. They were very pleased about our enthusiasm for the garden, which they had feared might fall into the hands of 'philistines'.

So, our venture into what had first appeared to be a lonely hamlet was blessed with interesting and friendly neighbours, but it didn't really blossom until we met the Morgans. We were not devout churchgoers, but we did occasionally attend the local church at Kirkby Overblow, where Sue was baptised into the Church of England. It was the vicar who introduced us to Trevor Morgan, who was busy converting an old farmhouse about half a mile away from us and was living rough until it was fit to receive his family. He was happy to accept Margery's invitation to Sunday lunches and proved to be an entertaining visitor.

Trevor's knowledge of the footwear industry was phenomenal, especially with respect to Italy, where he went frequently as a wholesale buyer. When his attractive wife, Benita arrived with their three sons, (and another on the way) the acquaintance developed into a strong friendship, for Richard and Martin were about Sue's age, and they got along splendidly. We soon started taking the boys out on picnics because they proved to be ideal companions—bright, cheerful, and brimming with fresh ideas, to such an extent that we couldn't think of taking Sue on our holidays without them, although we really couldn't afford such extravagance. So, we stretched our resources to the limit, and our annual holidays to Brancaster and Overy Staithe in Norfolk were memorable and worth every sacrifice we made. We sailed model boats, flew model aeroplanes and gliders and used the beach and seaside to capacity, including dinghy sailing. When it was too wet to go outdoors, the boys suggested plastic modelling, then in its infancy. Our first model was a Bentley, and I made an awful mess of it, because what I took to be glue was really a solvent that dissolved the parts that were being joined, and in fact welded them together. The more liberally the 'glue' was applied, the more the parts became distorted, so our first endeavours were disappointing. The boys didn't mind slight imperfections as long as the wheels went round, and we progressed to making

Chapter 6: 1955–1962 Pasture Cottagee

Sue and the Morgan boys, who turned out to be great companions.

aircraft with propellers that spun marvellously when held out of the window of our car—until their shafts melted with the friction and the props fell off!

Our enthusiasm increased to a point at which the boys even prayed for rain, and modelling became such an obsession for me that, when we returned home, I elected to make models while Margery watched TV, and I continued to make increasingly refined model aeroplanes for the next fifty years. Martin still feeds me with technical information concerning aircraft and recently sent some photos he took at the vintage aircraft museum in Moscow. He, and his family, remain good friends with Sue and her children.

Once Sue had developed friendships at Gateways School in Harewood, we became embroiled in birthday and Christmas parties; events I had carefully avoided from early childhood, and now faced with apprehension, because they involved social niceties of which I had no experience. One Saturday, Sue was whisked off by Helen Crockatt to a party at the Burtons, and I was told that all I had to do was to collect her afterwards. I was happily working away in the garden when Margery realised that I should have been at the Burtons half an hour ago. I had been doing some heavy digging and badly needed a bath, (we had no shower) but there was no time for that. So I had a quick wash, put on an overcoat to cover my disreputable and dirty gardening clothes, and promised Margery that I wouldn't even get out of the car. I had the Burton's address, but I wasted a long time finding it, for I was searching around for a normal house, whereas it turned out to be a 'stately home' in its own extensive grounds. Of course, I might have known, for the Burton family were very 'well heeled', owning over 200 shops throughout England known as the 'Fifty Shilling Tailors', where they sold 'ready-made' suits from their Leeds factory for two pounds and ten shillings—a competitive price, even in those days. Stanley Burton was a member of the University Council, and a generous benefactor to the university, and now I was to meet his brother, who evidently lived in some style.

Being so late, I expected to find a tearful Susan waiting on the doorstep for me. But no, the drive was so full of expensive cars that I had to get out and go up to the opulent entrance hall where I found a crowd of beautifully dressed mothers and fathers and their progeny, chatting away over drinks. To my horror, Mrs Burton insisted on my taking off my overcoat, going inside and having a sherry. Trevor Morgan would have made a joke of my predicament and made people laugh, but in my gardening clothes I was overcome by shame and embarrassment. I imagine the closest the Burtons ever came to gardening was to pick a few roses, or maybe prune them. So I made my apologies, collected my daughter, and we pressed our way through the crowd as speedily as we could, but not before I overheard one little girl say to another, 'I say—is that Susan's daddy? Isn't he *horrid!*'

Poor Sue. Whereas I was merely embarrassed, she must have been mortified. I never did discover how this affected her social standing, although, of course, when we eventually got round to holding a party, we gave it everything we had, with impressive catering from Margery. I have only a hazy recollection of the entertainment I cobbled together, but I do remember my desperate efforts to control one very naughty little boy who tried to sabotage all my ideas, and whose name is engraved in my memory; 'Mark Burton'.

All Susan's friends at Gateways seemed to have wealthy parents, and one friendly couple insisted on taking us out to a meal at an expensive restaurant in the country, where we felt a bit 'out of it' until Professor Dainton appeared. He took me to one side and whispered, 'What do you think of this place, Geoffrey? I couldn't afford it—I assume you're a guest like me?' This was typical of Fred Dainton's modesty. No one would have guessed that he was an eminent professor of Physical Chemistry who had just arrived from Cambridge with a team of researchers, for whom I had arranged special accommodation for their radioactive Cobalt Source. This was the start of his dazzling career, which subsequently included appointments such as Vice Chancellor of Nottingham University, Chancellor of Sheffield University, and Chairman of the UGC, Fellow of the Royal Society, and Chairman of heaven knows how many Government Committees. He was ultimately made a Baron and became greatly respected in the House of Lords as a man whose devastating arguments did not cause offense because of his exceptionable tact and irresistible charm. Anyway, we enjoyed a lovely meal, with a surprise afterwards when our host made his apologies on the way home, explaining that he had to make a detour to call at his 'places in Leeds'. These turned out to be a couple of small snack bars in the centre of town. He always collected the night's takings before going to bed; the income he received from these two insignificant little establishments was incredible—he was well on the way to becoming a millionaire.

Our cottage was about two miles from the bus stop on the Leeds/Harrogate Road, where I caught the bus to go to work. It was a pleasant half hour's walk on a bright summer's morning but became a long way home in bad weather at night. I would call Margery to tell her what bus I hoped to catch, and she would wait with the car at the end of the lane. There were times when trains were late, or the bus service was delayed by fog, so it was an imperfect arrangement for her, having a baby daughter to care for as well and no mobile phones in those days.

Eventually, I found myself bedridden, trying to get rid of a duodenal ulcer, when Margery walked in and announced that she had bought a car! I sat up, dumbfounded and was only mollified when she argued that the daily bus journey to Leeds was 'just too much'. She had bought, with her own savings, one of the new 'Austin Minis'—the very first of its kind ever to be seen in the Leeds district! We found it a delight to drive, economical on petrol and admired by everybody as the smartest thing around. The high opinion I already held of my wife increased by several notches and I looked forward to the day when I could own a mini of my own. Having two cars transformed our lives, and she was right about the stress; when a surgeon opened me up some years later to remove a gallstone and my appendix, he counted the scars of seven duodenal ulcers.

Living at Pasture Cottage was idyllic in the summer, but we wondered how we should fare in snow. We needn't have worried, because farmers fitted snow-blades to their tractors and cleared the roads early each morning for the lorries that were vitally necessary to collect their milk. It was when I reached Leeds that snow slowed me down. That changed one night when I had to attend an evening meeting and was late going home. There had been a little fine snow during the day, and the main roads were clear, but when I turned into Kearby Lane, I found that fields on either side had been swept clean by the wind, and the snow neatly deposited in the lane between the hedges. I was finally stopped by a drift some two feet deep, got out my snow shovel, and started digging. It took about an hour to clear enough to squeeze through to the lane beyond, which the wind had cleared beautifully, but as I got to the bottom of the hill and turned the corner I found the road completely blocked as far as I could see. So, I gave up and backed the car into a nice open space completely clear of snow. The digging had left me utterly exhausted, and the walk home took till about midnight. Early next morning, I walked down the lane intending to drive to work. But the car had disappeared; a change of wind had completely buried it in an enormous drift. So I had to walk to the bus stop and wait another day before I could get it out, thankful that Margery and Sue had been away at the time. I expected some expressions of sympathy for my ordeal—but both of them thought it hilariously funny.

During winter, we fed wild birds nesting about the house and garden, but owls were not so lucky. In the bitter cold, they fed on carcasses of animals lying on the main Harrogate road, only to be hit themselves by the heavy traffic; I was not very popular whenever I interrupted the flow of commuters' cars to pick up a wounded bird. Most were beyond any help and died in the night, but one beautiful barn owl survived and

Margery took it to the vet for its broken wing to be set most successfully. Our mousetraps were kept busy feeding it for several days, and it was such a joy to see it eventually fly away.

And so, we spent six happy years at Pasture Cottage until the field next to us was advertised for housing, and it was clear that this was the sort of development that would destroy the rural character of Kearby that we loved so much. We agreed that we might do better elsewhere, and Margery undertook the searching. Housing was still in short supply, but at last she came upon an advertisement for a farmhouse in Summerbridge, approached by a cart track from the main Pateley Bridge road. An old man tending his chickens at the bottom of the track warned us not to attempt to drive up in our car because the track was 'very rough.' He wasn't exaggerating. He added, 'An' when tha's bin up theer once, tha'll niver go up agin!' So, we parked the car, and walked the half-mile up to the empty, dilapidated, lonely, lovely old farmhouse that was to be our inspiration for the next thirty years.

Chapter Seven

1960–1978: The Chamberlin Plan

Our extensive search for an architect and planner to replace Dr Lodge, and subsequent developments at Leeds, are described in detail by Professor Whyte of Oxford University in his research article 'The Modernist Moment at the University of Leeds, 1957–1977'[1]. This admirable paper makes fascinating reading and covers some interesting correspondence that I had quite forgotten, but I do remember the bare essentials of this exciting period of my life. It started with the selection of four promising young architects; Denys Lasdun, James Stirling, Bill Howell, and Peter Chamberlin, who were asked to spend a day with me examining our problems. The Vice Chancellor had said, as I would be working with one of them for the next ten or twenty years, he wanted my assessment of their personalities and capabilities before they met the committee.

The selection proved to be much easier than I anticipated; the first interviewee didn't bother to come to see me at all, but merely turned up for the dinner, when he produced a development plan based on maps he had found in a London library that were hopelessly out of date. His drawings did not show any of our post-war buildings, and were based on a pre-war city zoning plan, so were a complete waste of time. The next two were quite impressive, and the Chairman and I

---

[1] The Cambridge Historical Journal/Volume 51/issue 01/March 2008, pp.169–193

accepted invitations to inspect their work at Leicester University and Ham Common. They had put up exciting buildings at other universities, but they both struck me as personalities who might be difficult to work with, and I was very relieved when the time came for us to interview the fourth, and last, nominee, Peter Chamberlin.

In contrast with the other three, Chamberlin spent a whole day examining everything I could show him with genuine interest and courtesy, and I reported to the VC that this was someone who I would be very happy to deal with. At the committee meeting after dinner, he showed us a development plan that had been commissioned by the Corporation of the City of London, covering seven acres of bomb damage around the ancient Roman remains known as the Barbican. It included a school for music and drama, a library, art galleries, two theatres, three cinemas, seven conference suites and two exhibition halls, extensive housing, and shops with imaginative landscaping throughout. It was a comprehensive report bound in hardback and was very impressive indeed.

Chamberlin's partners were Geoffrey Powell, who had won the Golden Lane Housing competition in 1951, and Christoph Bon. All three had been teachers at the Kingston Polytechnic, and their partnership became acknowledged as 'one of the most important modernist firms in post-war Britain', with many of their works becoming listed buildings. One of these is the elegant New Hall, Cambridge, completed in 1964. It became clear that Chamberlin was not so much a 'promising' architect but one who was 'flourishing', for he was in the enviable position of being able to accept only such commissions as appealed to him. It was fortunate that he judged our university to be one of them. The committee did not have any doubts about his appointment although they were taken aback somewhat by his insistence that he would need twelve months to do research on our problem before he could produce a satisfactory plan.

As to Chamberlin himself, who was affectionately known as 'Joe', I soon came to have a great respect for his imagination, intellect, and modesty, as did most of the members of the university who met him. On his first visit to Leeds, he said that he had seen enough of the campus for the moment, and could we retire somewhere quiet and get to know each other personally? It was a lovely summer's day, and as we were living at Pasture Cottage, he came home with me and we spent the afternoon talking and taking tea in the shade of an old apple tree. His conversation was informative and stimulating, for his interests ranged from modern art to classical music and even to risqué pop songs from America. After tea, he entranced our young daughter by his ability at origami—folding a large beak out of cardboard that made an audible snap and which he called a 'cockie ollie bird'.

At the time, I was apprehensive about the fees he might be charging for a visit such as this, but I needn't have worried, for it transpired that all three members of his partnership were so well endowed financially that they worked for the sheer love of creating fine architecture. His fees were always related to results rather than to time, and expenses were always very reasonable. I was to discover, much later on, that his house in Chelsea was crammed with the very latest hi-tech equipment for cooking and entertainment and, as early as 1965, he was looking forward to hanging a plasma television set on his wall. He grew peaches and grapes at his island house on the Thames and took holidays in his villa in the South of France.

Sir Charles Morris often invited Joe to dinner at the Vice Chancellor's Lodge for informal discussions, for he liked Chamberlin's personality and enjoyed talking with him, especially about some of his original ideas that provoked debate in the Senate. He found Joe convincing in his arguments, based on his philosophy 'that nothing less than the best was good enough for him or his clients'.

By 1960, after more than twelve month's research, the Chamberlin Development Plan was published in the form of an impressive book with photographs, drawings, and diagrams. It had entailed a prodigious amount of labour, not only by the architects, but also by our Planning and Registrar offices, which had been required to survey every tree in the projected precinct, to list every room in all our existing buildings, and to prepare statistics of student numbers, existing and projected, timetables and details of the courses they took, so that their movement between buildings could be plotted by the architectural planning team.

The result was an extraordinary series of diagrams that illustrated how the university operated in physical terms, and how this mechanism was affected by the location of its buildings. They also highlighted the problems of student movement between classrooms and laboratories. Long tradition held that classes should finish at five minutes to the hour, and start at five minutes past, and it was obvious that as the campus expanded it would be difficult to keep to this 'ten-minute rule'.

Chamberlin pointed out that more time was spent in going up and down staircases and getting in and out of clothes in bad weather than actually moving between buildings. He suggested continuous covered walkways at different levels; these would shorten the time spent on staircases and also reduce the number of cloakrooms necessary. Each of the walkways would have a different colour, and eventually the elevated Red Route became a dramatic element and one of the best-known features of the university. In this way, tradition was upheld and the Registrar didn't have to amend any of his timetables.

In discussing the design of departmental buildings with Joe, I had laid great stress on the need for flexibility. Student numbers were constantly changing, and new professors could bring with them research teams and equipment, such as electron microscopes or even radioactive isotopes that demanded elaborate air conditioning and so on. Such facilities were fiendishly difficult and expensive to incorporate into traditional buildings. Chamberlin's response was to plan departmental buildings in the form of continuous spines that enclosed quadrangles and created open spaces for gardens reminiscent of Oxbridge colleges. Committee members were surprised at my waxing lyrical about the brilliant design of these spines that appeared to them to be no more than dull, monotonous buildings, but, to me, they held the promise of a solution to all my difficulties.

Rooms of any size could be created simply and economically by partitions, unhindered by the location of windows, because fenestration was continuous along the whole length of the building. Lifts and staircase could be installed almost anywhere without structural modifications, and pipework and ductwork could be installed in the space between columns and parallel beams. My own design for the Engineering Block had made a small step in this direction, but this was an inspired design that took things way beyond anything I could have conceived. I was thrilled, because, to me, it was beautiful, and in practice it proved to be even more successful than I could have hoped.

When it came to lecture theatres, I anticipated real trouble, for it was an established tradition that every department should have its own theatre, big enough to hold all its students and staff for lectures and staff meetings. It was usually the first room to be shown to any visitor, its size being a status symbol that reflected the department's importance. I knew that most of these larger theatres were used on average only a few times a week, and only during term. They were very expensive to build, and if they were tiered had useless space underneath them. So I was delighted when Chamberlin produced a very clever solution with twenty-five theatres stacked together in a neat structural 'package' that had a remorseless logic to justify its design.

Our traditional theatres had doors through which up to 150 students had to thread their way after a lecture, and then had to push their way through crowds waiting to get in to the next lecture. This congestion was endemic and threatened the ten-minute rule. Chamberlin's solution eliminated this by having a door at the end of each row of seats, opening directly to the staircase outside. As a result, only fifteen or so would be going out of each door after a lecture, distributed evenly over the staircase, where fifteen students would be waiting to get in. It was beautifully simple and saved money by eliminating the need for lobbies and internal aisles and making it possible for the benching to form part of the structural floor of each theatre.

The arrangement of stacked lecture theatres also eliminated the wasted space under the tiers and secured other significant savings, such as requiring only one air-conditioning plant instead of several. Another innovation was the provision of TV monitors in every theatre, for it was anticipated that lectures in the future would be recorded—at least those given by eminent speakers or those involving elaborate scientific demonstrations. With this in mind, a central TV studio was planned in the basement with appropriate laboratory facilities. At that time, black and white TV was in its infancy, and colour TV was a pipe dream. I well remember Mr Holroyd, the Director of the service proudly showing me one of the very first black and white reel-to-reel TV recording machines that he had persuaded the BBC to let him have second hand for the bargain price of £10,000. It was huge and required two men to lift it. Many were sceptical about his new department, and it certainly had teething troubles, but, eventually, this embryonic idea blossomed into a flourishing Institute of Communication Studies.

The 1960 Development Plan was printed as a large hard-back book similar to the Barbican Report, and the arguments and explanations were set out so clearly that most members of the university accepted its proposals with enthusiasm. Copies were sent to every British university, to important universities abroad and also to the Hospital Board and the city engineer. These last two copies 'set the cat among the pigeons', resulting in reactions that were unexpectedly hostile.

The city engineer's pre-war plan for the city centre had incorporated an inner ring road intended to reduce the amount of traffic though the heart of the city. It was routed between the hospital and the university. We had taken it for granted that the old 19th century infirmary would be rebuilt, so our brief for Chamberlin had called for a new Medical and Dental School to be closely coordinated with any hospital developments that might be proposed in the future. Chamberlin pointed out that a busy roadway on the surface would create serious noise and pollution problems for both the hospital and the Medical School and he proposed that the ring road should be sunk in a tunnel. This would allow new buildings for pathological sciences, medicine and dentistry to be built over it to efficiently link the university to the hospital.

It was this revolutionary concept of a sunken ring road that aroused such fierce opposition from the Hospital Board and the city engineer, but our enthusiastic university was determined to fight for what was clearly the best practical solution to the problem. The real reason for the hospital's objection became apparent when we learned that an extension then being built for the infirmary was to be named after the chairman, Sir George Martin. He probably thought that Chamberlin's proposals would detract from the importance of the Martin Wing and was not slow to take action. He arranged for the Yorkshire Post to publish a drawing showing how new hospital buildings could be linked to the university by a bridge, without the 'extravagance' of a tunnel. The author of this pathetic proposal was a Mr John Poulson.

Sir George went even further, suggesting that Mr Poulson might be appointed architect for the new Medical and Dental Schools, so the Vice Chancellor arranged for me to meet him. After I had declined his pressing invitation to dine at the Queen's Hotel, we went to the Planning Office and discussed architecture in general. I found myself having to listen to a lengthy recital of Poulson's achievements, his numerous commissions, and above all, his personal success in a way that made me realise that he was far more interested in making money than in creating fine architecture. I reported to the VC that I could not understand how any responsible person could dream of employing such a man for a major project.

I later discovered that Poulson was not a qualified architect, for in 1927 he had dropped out of the architecture course at the Leeds Polytechnic to work for a firm of architects in Pontefract, who sacked him in 1932 'because he tended to get elevations ... the wrong way round'. Poulson then set up his own business and, because he was not a Registered Architect, simply ignored the Professional Code of Conduct requiring bills of quantity and competitive tendering procedures that were designed to protect the interests of the client. Instead, he employed his own building workers, and quoted for the work himself, persuading clients that this was far more efficient than going through 'old fashioned tendering procedures'. Of course, as variations to the contract became necessary, the absence of a bill of quantities made it impossible for anyone to check the validity of his extra charges. The fact that Poulson had secured so many prestigious contracts puzzled me until I learned of these business methods that would have been quite unacceptable to any businessman spending his own money. However, in the case of organisations disbursing public money, his arguments were accompanied by offers of personal inducements to committee members if they agreed to take advantage of his unique services.

At the time I gave him his interview he was building a large, unattractive office block near Leeds City Station, and in 1966 he completed an International Swimming Pool (which the city engineer later found to be leaking). He employed around 750 people, in four offices in England, Lagos, and Beirut. He was a millionaire but was reported to be an uncaring employer who sacked assistants, usually just before

Christmas, for trivial offences such as growing a beard. In 1968 his empire began to collapse with the Inland Revenue successfully suing him—although he was himself a Commissioner for Taxes. There followed a series of massive corruption scandals that became known as 'the Poulson affair', with investigations for fraud, a jail sentence of seven years for him and smaller ones for several town councillors. Three Members of Parliament escaped through legal loopholes, but Reginald Maudling, the Home Secretary, became hopelessly entangled in the web of corruption and had to resign in disgrace. Afterwards, I was relieved to think that by declining his invitation to a meal I had avoided any hint of contamination. Nothing ever came of the Poulson Plan, and, in due course, Sir Donald Kaberry MP became Chairman of the Hospital Board and joined the university in appointing a Building Design Partnership as architects for a new hospital together with our new medical school.

The objection from the City Engineer, David Currie, was more substantial and understandable. After months of tedious negotiation he had at last succeeded in persuading the Ministry of Transport to fund his inner ring road project, but they certainly would not pay the extra cost for a tunnel. Our approach to the University Grants Committee met with a similar response; their funds were 'for putting up buildings—not for digging underground roads'. Eventually, Brigadier Tetley suggested that money already given by the Tetley Brewery to the University for 'general purposes' might be used to provide the £700,000 needed.

We were then relieved to find that, following Mr Currie's recent retirement, his young successor, Geoffrey Thirlwall had found our proposals to be a heaven-sent opportunity to revise the old plan for surface roads and roundabouts that he regarded as obsolete and quite unsuitable for post-war traffic. He not only embraced the concept of a tunnel, but also extended the idea of a sunken ring road to encircle the northern half of the City Centre, with radial roads using bridges instead of roundabouts. To the delight of everybody, the Ministry not only promised the extra funding, but also displayed Thirlwall's drawings in their London Headquarters, extolling them as a model for other cities to follow. We agreed with the city that, as we were paying for it, the university's consulting engineer, Dr Anthony Flint should design the tunnel and our clerks of

Chapter 7: 1960–1978 The Chamberlin Plan

works would supervise it, while the city would be responsible for the roadway through it (when we first appointed Dr Flint, he was a senior lecturer at Imperial College London and was just starting his professional practice; he eventually received the Gold Medal from the Institute of Structural Engineers, and later became their President).

After two years, the tunnel was complete, and during one of my routine meetings with the City Engineer it was mentioned that the Minister of Transport was coming to Leeds to perform an official opening ceremony. 'And who is going to represent the University?' I asked. It was shamefacedly admitted that the fact that the university had initiated the idea of the tunnel, and had designed, supervised, and paid for the whole thing had been completely forgotten. I don't know who attended the ceremony, but I took care it wasn't me, for it was to be on a Saturday, and my weekends were very precious, for at that time I was busy building a barn at our new home at Woolwich Farm

A welcome side-effect of the city's road programme was the demolition of dozens of houses; a great help to the university in its laborious acquisition of land, some of which was earmarked for our residential use. 'But, what type of residence', asked Chamberlin, 'had the university in mind?' It was a very good question, for with the social and sexual revolution of the 1960s, students were demanding more freedom in their living accommodation, and flats were being mooted, because they promised to be cheaper than halls of residence to build and operate.

Being a university, the obvious answer was to do some research and I was sent, with the Bursar and Dr Belton, on a fact-finding tour of European universities, where student flats were the norm. Starting in Finland, we toured through Sweden, Norway, Denmark, Germany, France, and Italy. I reproduced plans and took photos of every residence we visited, dictating descriptions of them, to be posted to my office the next day and typed ready for our return. It was arduous work, and in the meantime, Joe Belton and Edmund Williamson enjoyed some sumptuous evening meals, while I (because of my ulcers) had to subsist mainly on Complan. However, I did seize the opportunity, in every country we visited, to make a 'scientific' study of Peche Melbas, having tasted one in Helsinki so scrumptious that I suspected that it could not be bettered—and I was proved right.

Indeed, the choice of Finland as a starting point of our survey was most propitious for it was a revelation to us, not only in terms of sophisticated architecture, but also in social behaviour. At that time, British students tended to dress casually—one might almost say sloppily, to assert their sense of personal freedom and support of the revolutionary student movements then surfacing in Europe and America. This tendency could, and did, result in clashes between social norms in England. For example, it was some years later, when Sir Roger Stevens had succeeded Sir Charles Morris as Vice Chancellor that, in keeping with this atmosphere of mutual cooperation, he took the unprecedented step of inviting the student President to propose the vote of thanks to the Chancellor, the Duchess of Kent, at the conclusion of one of the Honorary Degree Ceremonies. Of course, this was a full-dress occasion held in the Great Hall with all the staff and the Lord Mayor of Leeds and mayors of several other towns in their full regalia. The President, (having borne the Mace in front of the Chancellor in the opening procession), bowed to her and begged permission to ask one of the Student Council to propose the vote on behalf of the students. Her Royal Highness graciously consented, (what else could she do?) and the President then called upon a Mr Quail, who was in the front row of the gallery, to speak.

His speech began well enough, but gradually deteriorated and became a tirade against landed gentry, their abuse of privileges, and so on. The embarrassed students behind Mr Quail vigorously stamped their feet in protest, and staff on the platform around me started growling, 'how long will the VC allow this outrage to continue?' But Sir Roger, who had been a career diplomat, and Ambassador to Persia, did nothing to stop the speech. When he rose, he solemnly thanked Mr Quail and went on to explain to a shocked audience that he could think of no better illustration of the value that this university placed upon the freedom of speech than was manifested by Mr Quail's address. With thunderous applause, the congregation rose and the President, after bowing to the Chancellor, led the procession from the platform. Sir Roger was boiling with suppressed anger and embarrassment as we walked behind him along the Red Route on the way to lunch. Eventually he turned to the Registrar, Dr Loach, and said 'John, isn't there a season for shooting quails?'

Chapter 7: 1960–1978 The Chamberlin Plan

It was against this background that we arrived at the Student's Union Building in Helsinki, to find a refined atmosphere that would have done credit to any Gentleman's Club in London. Every man had the traditional fur hat and overcoat, for it was February, and these were deposited at the manned cloakroom near the entrance. Most students wore suits, shirts, and ties, and the women dressed with similar decorum. The subdued noise level was more like a library than a students' union. I didn't see any posters apart from some small ones in the lobby where notices were displayed. The Student President greeted us warmly in perfect English and took us to the comfortable coffee lounge. He was an engaging young man with a wonderful sense of humour. He told us that he was once showing a German visitor the view from the Northern frontier, when they joined an elderly woman sitting on a bench. As soon as she discovered that the visitor was German, she abruptly got up and stalked away in high dudgeon. Our friend apologised for her rudeness and explained that most of the properties along the frontier had been destroyed during Germany's war with Russia. 'That's quite understandable', said the visitor, 'if her house was demolished'. 'Oh, no', said our friend, 'she's furious with Germans because they didn't touch her house, so that, while all her neighbours got compensation and rebuilt their homes, she got nothing!'

He had taken a sabbatical year to serve as student President, being provided with free accommodation, a salary, and a car, and told us that students regarded themselves as privileged to be at university, having real power, owning and managing their Student Union Building without any help or advice from the university. This was in sharp contrast to Leeds, where the university owned and maintained the union building, managing it through a rather clumsy advisory committee, consisting mainly of students, with Dr Belton as Chairman and myself as Secretary. He impressed us further when he showed us the large commercial block that the union owned in the city, which provided a significant income for them, and took us to see residences built with their own funds that ran at a profit. Indeed, when the university decided to move further from the centre of Helsinki, some student housing was completed on the new site before any of the teaching accommodation was ready.

All this began to make sense when he pointed out that the average age of students was twenty-two, and most had completed their military service. Alumni were keen to help, providing a professional service of lawyers, architects, engineers, and businessmen at little or no cost. A similar tradition was held in Stockholm, where the chairman of the student building committee was a past President of the Swedish Academy of Architects.

In Finland and Sweden, rooms were allocated to men and women indiscriminately, their social and moral behaviour being regarded as a matter for their own discretion as they were deemed old enough to behave like responsible adults. This was also the case in Norway and Denmark, and when we asked a high dignitary of the University of Copenhagen to comment on a widely publicised incident of immorality in an Oxford College, he said, 'In Denmark, we would not have looked'. What was even more fascinating, was the discovery that this attitude towards sexual behaviour and maturity steadily changed as we travelled south. It reached a point in Italy where no student, not even of the same sex, was allowed to visit another's room, and a concierge was on constant duty at the junction of the corridors to make sure that no one did. When we asked our two guides, aged twenty-one and twenty-two, 'When do you mix with the boys?', they giggled like young schoolgirls and assured us that they were allowed to speak to boys, under supervision, in the common room. In the dining room, where they had to sit at different tables, they threw buns at each other. This was in 1962!

While we were in Stockholm, I showed my colleagues the Southern Hospital, which I had first seen during an RIBA tour of Scandinavia in 1947 when it had been the biggest hospital in Europe. The Swedes had been very proud of its medical innovations and equipment but admitted to a complete catastrophe in the catering arrangements (my military friends in Toc H would have characterised this as a monumental cock-up). From the earliest planning stage they realised that the sheer size of the hospital would exacerbate the problem of conveying freshly prepared food to patients without it getting cold. So the design team had decided to tackle this problem head on by selecting the brightest graduate from the Stockholm College of Catering, sending her for two

Chapter 7: 1960–1978 The Chamberlin Plan

years to a technical college to be taught by specialists in the design of kitchens, and then setting her to work to find a solution. They were very pleased with the results, and were convinced that they had a good system, in which the main bulk of the cooking would be done in a large central kitchen, but the individual meals would be finished and served from small kitchenettes on each floor immediately adjacent to the wards.

The whole of this huge Swedish Hospital had been built on this principle, but when the Management Committee took over and interviewed applicants for the post of Catering Manager, they concluded that the young lady who had conceived the whole system was 'too young for such an important position'. They gave it instead to an elderly woman with many years' experience in smaller hospitals, all of which had used traditional systems. The design team was then astounded to hear that the new manageress had, without the common sense to give the system a trial, persuaded the committee that it wouldn't work. They were now enraged to hear that the small kitchenettes were to be redesigned as serveries and the main kitchen in the basement was being enlarged in spite of the considerable expense involved. They had no idea when the hospital would be opened, nor what standard of catering would be achieved in spite of all their foresight and effort. This 1947 architectural tour was intended to bring British architects, who had been unable to design anything for six years because of the war, 'up-to-date' with the latest developments in Europe, especially in regard to high-rise flats, which they thought might help with our housing crisis. We were shocked to learn that the Swedes were worried about increasing numbers of suicides in their flats, which they blamed on loneliness. In fact, they said, at that very moment a team of their planners was touring English housing estates to study our semidetached housing, new to the Swedes. Many years later, in 1962, Chamberlin showed that the idea that high rise flats saved space was a fallacy, for they required so much breathing space that the same density per acre could be achieved with housing of only three or four storeys.

When we moved on to Norway, we found that conditions were dramatically different. During the war, the Norwegians' heroic resistance to German occupation had done nothing to improve their prosperity, whereas the Swedes' neutrality had enabled them to profit from the war. On the 1947 tour, our guide had proudly pointed out a small ship in Stockholm Harbour, which, he said, had regularly run through the German blockade to get steel ball bearings over the North Sea to England. One cynic in our party later commented, 'he didn't show us its sister ship that ran the British blockade to get ball bearings to Germany' (neither of them knew that, in fact, we had used unarmed mosquito aircraft for these precious cargos). Consequently, Oslo had few new buildings to show us. After the war. the university had found that the only way they could cope with the increasing demand for higher education was to have two universities sharing the same buildings, using a shift system. The morning shift was from 6.00 am to 1.00 pm. The afternoon shift was from 1.00 pm to 8.00 pm. After that, the evening classes took over. Cleaning and maintenance had to be done in what remained of the day, from midnight to 6.00 am. This arrangement worked surprisingly well, for it seemed that there were two types of personalities; some students were quite willing to get up early and to have the afternoons free for sport, and others preferred to work late into the night and to sleep in the morning. I always had admired the Norwegians, and this was efficiency at its peak.

In Paris, student housing appeared to have been designed by some builder determined to cram as many rooms onto the site as he could, and at minimum cost, for the results we saw were quite dreadful. I reckoned that they might conceivably have stimulated the subsequent student riots, for the monotonous repetition of hundreds of identical rooms appeared to us inhuman in scale and overwhelmingly depressing.

There was little of note in Freiburg, or in Zurich, because Swiss universities traditionally leave students to arrange their own accommodation, but in Milan, there were striking differences from everything else we had seen so far. The style of architecture in the halls of residence resembled an Italian palace, with marble floors, wide corridors, high ceilings, and generous study bedrooms having a washbasin, but with a cold-water tap only. The absence of any carpets or rugs on the marble floors emphasised a clean but austere environment consistent with the extraordinary social restraints I have described. Milan provided us with a fitting anti-climax with which to complete our survey.

Chapter 7: 1960–1978 The Chamberlin Plan

As we passed by on our way to Switzerland we sneaked a visit to Le Corbusier's little Chapel at Ronchamp. Although I had always respected Corb's fresh ideas, they sometimes had a brutality that worried me, but here was a design with an impact like a breath of fresh air. True, the roof seemed to be heavy, but it floated above continuous clerestory glazing above the walls. These were unusually thick but were punctured by a variety of windows letting coloured light flood the chapel. Whereas traditional leaded lights in English churches tend to draw attention away from the inside of a church, the stained glass in the Ronchamp chapel contributed to the warmth and interest inside the auditorium, and the thickness of the walls gave a sense of privacy and safety. Everything about that building was a delight; it was truly inspirational.

Back home, with my schedule of accommodation based on our detailed report, Chamberlin designed a block of flats to run alongside Clarendon Road. Each study bedroom was to share a tiny washroom with shower and WC and the bed was designed to double up as a settee. It was a very tight design, with original and ingenious features, and as several hundred identical rooms were to be built, it was important to get everything right first time. So, we built a prototype in the Parkinson basement not far from the planning office and we tested it remorselessly, taking turns to sleep in it, and suggesting several amendments in detail. These 'Henry Price' flats proved to be a landmark in British student accommodation, and attracted visitors from all over the world. Consequently, our European tour was judged to have been a success, and when Chamberlin designed the Charles Morris Hall in the centre of the campus he adapted a similar arrangement for the men's study bedrooms.

In the meantime, we had been considering the pressure for student housing that continued to be relentless, and, since all available sites had now been developed, it became clear that we should have to build on some of our land at Weetwood that was not suitable for conversion into more playing fields. There was more than enough room for at least four halls of residence, each housing 150 students, but eyebrows were raised when I put forward the relatively young Denis Mason Jones as architect for the whole project. I had enormous faith in his capabilities; his designs were elegant, sensitive, and thoroughly practical. His father had designed buildings for the university before the war, and Denis had a degree in architecture from Cambridge. He was also exceptional in having worked on site as a bricklayer to learn the practical side of building. What was particularly endearing to me was his tight control of costs, for which I was ultimately responsible, for I had directed some dozen architects and consultants by this time and keeping their expenditure within bounds was a constant nightmare. So, I regarded Denis as something like an angel.

The figure of 150 was based on economic and social factors, and was generally accepted as a norm. My years of living in Devonshire Hall, with its full-time warden and 150 students, had been followed by a much happier period when I was the part-time warden at Woodsley Hall with only seventy-five students. The comparison was striking. At Devonshire Hall it was difficult to get to know more than a fraction of those in residence, whereas the seventy-five men at Woodsley constituted an ideal social group in which each student quickly came to know everyone else.

It occurred to me that instead of four halls of 150, we might have eight halls of seventy-five, each with its own part-time warden, and all sharing central facilities with a single steward and catering staff. This should not only be more economic but would avoid the troublesome problem of recruiting full-time wardens. Good ones were not easy to find, and bad ones could be very difficult to get rid of. The committee liked the idea of part-time wardens forming a council and sharing administrative duties, while being responsible for their own group of students, and my proposals were approved.

Denis duly produced an attractive plan, but it immediately ran into trouble when we applied for planning consent, for the city engineer was anxious to preserve a 'green belt' of open space around the city. He regarded our playing fields as an important element in this. We conceded that he had a point but argued that the site for our proposed new hall would use up a relatively tiny portion of the open land available, of which only about half was being used for recreation. We also reminded him that the city itself had only recently acquired from us, under threat of a compulsory purchase order, a portion of those same playing fields on which to build a police station.

The City Council might well have given way to our arguments, but householders whose property enjoyed extensive views over this open ground soon raised objections, partly about the loss of an amenity where they liked to exercise their dogs, but also because of the possible effect on the value of their houses. In view of their vigorous protests, the city had no alternative but to hold a Public Inquiry. They appointed a distinguished Queen's council, who specialized in town planning legislation to represent them.

The Bursar, Edmund Williamson, and I represented the university, and the QC immediately adopted tactics to make us both look silly. He asked why Sir Charles Morris had not appeared in person but had sent his 'head cook and bottle-washer' instead. He then proceeded to grill Denis very unpleasantly about every aspect of his scheme, and particularly about the capacity of the hall. In reply to one of his challenges, Denis compared the number of residents to the 600 living at King's College, Cambridge. 'And from whom did you learn the number in that college', he was asked.

'From the hall porter', replied Denis.

'The hall porter?' sneered the QC, 'surely you might have enquired from someone in authority—like the Bursar?' At this point, the inspector who was conducting the Inquiry, and who was evidently annoyed by the man's intolerable rudeness intervened, saying that, as it happened, he had been a student at King's College himself, and he could think of no more reliable authority on the number of students than the hall porter. It came as no surprise to us that the Inspector ruled in our favour.

I was delighted when John Woodhead, who had succeeded me as warden of Woodsley Hall, moved with his students to occupy the first of the new eight halls, which, as a group, were named after our first Vice Chancellor, Sir Nathan Bodington. One of the halls was named after my dear mentor, Professor Evans. With the opening of the hall, the Council of Wardens worked out better than I had dared to hope, with a lecturer in Economics looking after financial affairs, and my friend Bob Mackey, a lecturer in Civil Engineering very happily looking after maintenance. For this design, Denis was awarded the Leeds Gold Medal for Architecture from the Hoffman Wood Trust. In 1976, three years before I retired, the Student President of Woodsley Hall (and sometime later its warden) was a certain Christopher Snowden, who became Professor of Microwave Engineering in 1992, and Head of School in 1995. He was elected an FRS (Fellow of the Royal Society) and knighted and was Vice Chancellor of the Surrey and Southampton Universities in a dazzling career of invention and innovation. Somebody who Woodsley could be proud of, and I'm sorry I never knew him.

Bodington proved to be very successful and popular; further blocks in the form of student flats were built and numbers increased steadily from the original 600 to 1,140 by the time it reached its fiftieth anniversary, in the year 2011. However, sometime after that, it was reported that expensive refurbishment was required, and as students were showing a strong preference for accommodation nearer the main campus, it was decided that the hall should be closed and the buildings sold to a property developer.

This news came as a considerable shock to me, as can be imagined; I still regarded Bodington as new. I now realise that I must be older than I thought—but it was good while it lasted.

I never missed an opportunity to invite Denis to a function in the senior common room for people liked him so much and he was an unfailing source of good humour and wit. To the best of my memory, one of his delightful anecdotes went like this;

'Mary and I were coming home from a picnic with our three children and their nanny (we had to have a nanny because Mary is a full-time doctor). Mary always responds quickly to emergencies and tends to take charge of things. So, when she saw a whiff of smoke coming from under the dashboard she told me to get off the road, ordered everyone out of the car, and insisted that *all* our belongings must be taken out in case the car were to erupt in flames.

At the time, we were going along Harrogate Road, by the Harewood Estate and I parked on the very wide grass verge. I just couldn't believe how much stuff had been packed into that car, quite apart from all the paraphernalia we'd taken for the picnic. It was embarrassing; the place was beginning to look like a refugee camp. Being a Sunday made it worse, for the road was crowded with traffic returning from a day's outing. People in cars passing by were 'rubbernecking' to see what was

Chapter 7: 1960–1978 The Chamberlin Plan

happening. I couldn't see any more smoke and was just going to suggest that it might be safe to get back on the road, when Mary flagged down a passing car and got in. I knew she was attractive, but the ease with which she secured that lift came as quite a shock to me.

The next time I saw her, she was in a huge recovery truck coming from the direction of Leeds. It held up all the traffic as it made a U-turn to cross the road and took up a position behind our car. On board was the largest fire extinguisher I've ever seen. 'Where's the fire?' asked the garage hand. I started to explain but was interrupted by the clanging of a bell as a large fire engine swept down the road, followed by the Fire Officer's car. Once more all the traffic had to stop while both vehicles executed the U-turn and took their places behind the recovery truck. As the Fire Officer joined his men he asked, 'have you disconnected the disconnector?' 'No', said one of them 'you only hav'em on diesel engines'. It was while they were pondering what next to do that a siren heralded the arrival of a police patrol car complete with flashing lights. This did the obligatory U-turn and drew up behind the Fire Officer's car. The kids loved it!

All this time, the traffic along the main road was getting heavier and slowing down to a crawl to look at this incident, which must have looked very serious because it had called for every conceivable emergency vehicle except an ambulance. So, it was a relief when the policeman assumed control, sent the truck away and directed the traffic to move on. The firemen found no fire—just a short in the electrical wiring, and said it was safe for us to go home, apart from one snag; our rear lights weren't working. It was now getting dark, and the policeman said he couldn't allow the car on the road without rear lights, but he would arrange for the fire engine to escort us home.

So the fire officer and the police car went off, leaving us to pile everything back into the car, and drive home with the kids happily gazing through the rear window at the following fire engine. All went well until we had to slow down in to negotiate the narrow streets of our housing estate. As we got near home one of our neighbours ran towards us waving his arms excitedly and shouting, 'FOR GOODNESS' SAKE, DENIS, GET OUT OF THE WAY! THERE'S A FIRE ENGINE TRYING TO PASS YOU!'

These are the bare essentials of the tale is I remember them, but Denis embellished them with such witty comments that he always kept his listeners convulsed with laughter. He was a lovely man.

About this time, the erection of the first of Chamberlin's spine buildings to house the departments of Mathematics and Earth Sciences had reached roof level and a traditional 'topping out' ceremony was held, attended by Peter Chamberlin and me with due publicity in the local press.

Peter Chamberlin and me at the 'topping out' ceremony for the first spine building.

This building was to be linked to the much larger Physics building by a bridge forming part of the Red Route that started at the University Road level near the Arts Block, passed through block 19, went along the side of the library and through the full length of the Mathematics building. From here, via bridges two-storeys high, it went through the very heart of the lecture theatre block and joined the biological sciences departments along their top floor, ultimately reaching the new Medical Centre at the southern edge of the campus. It used to be called 'the longest corridor in Europe'; this no longer true, but it certainly impresses visitors. It achieves all that Chamberlin hoped for—comfort, efficiency, and security. It welds together the centre of the campus in remarkable harmony and comfort, having dramatic vistas of gardens and courtyards all along the route, with protection from the weather, and is regarded with affection by both staff and students.

The original concept for the library was to have a crescent shape, of which the roof would ultimately provide tiered seating for a gigantic open-air auditorium suitable for large degree ceremonies, enclosed at

Chapter 7: 1960–1978 The Chamberlin Plan

Top photo: The Roger Stevens Lecture Theatre block.
(Photo by Tom De Gay, courtesy of Avanti Architects)
Bottom photo: A side view showing the tiered lecture theatres
(Photo Wikipedia Commons)

---

the sides by spine buildings, and at the east end by some form of stage. It was to have a removable canvas tent roof on the lines of the Colosseum in Rome.

At the time, there was no large auditorium in Leeds, although the city was negotiating with potential donors to build a large city concert hall. Two wealthy men had offered to pay for half the cost of a hall, as long as it was named after them—but they couldn't be persuaded to

Bridges on the Red Route linked the spine buildings. The Maths and Physics Departments are shown here.

put their donations together. For practical reasons Chamberlin had to abandon this grandiose idea for a Congregation Court, and the shape of the library became a rectangle instead of a crescent. Even so, its final design involved months of arduous negotiations, during which Chamberlin submitted thirteen fully worked out designs, each of which was rejected, until the librarian, Dennis Cox, and I began to despair. All the proposals had attractive features, and Frank Woods presented them with persuasive arguments, but they failed to meet our criticisms, most of which related to the daily re-shelving of large numbers of books, and the movement of trolleys. Each solution was an improvement on the last, and, finally, Dennis expressed an exhausted satisfaction with a design that owed as much to his tenacity as to the architect's persistent ingenuity.

Although Christoph Bon and Geoffrey Powell contributed to several other projects in Leeds, such as the development of Woodsley Hall, into a residential college for Extra Mural Studies, they did not have the same charisma as Joe. As the number of visits to Leeds increased,

Chapter 7: 1960–1978 The Chamberlin Plan

Joe began to rely on Frank Woods, whose intellect and charm dealt smoothly with problems that arose in meetings in the Planning Office, or in Committees, or with the contractor. With so many major projects on hand there was no shortage of problems, but there was only one serious disagreement to record. The University Grants Committee in London had earmarked funds for a spine building for the food and leather industries, and Joe's design went way beyond the size and cost that they could approve. I pointed out how we might modify his proposals, strictly in keeping with his ideas, to fit within the budget, but Joe was obstinately reluctant to reduce the size of the building, until time ran out and we missed the grant altogether. With hindsight, I wonder whether he might have been seriously ill at that time.

In my view, Joe Chamberlin was an intellectual giant—but such a gentle giant, for his firm resolve was never overbearing—merely politely persistent. Whereas most architects conceived their designs with a view to erecting monuments to themselves, he resisted the temptation to make beautiful buildings alone but invested our money in such a way that the university would benefit every time it had to rearrange its accommodation. He was a man of integrity.

As a result of my experience with the Parkinson Building early on, I had instituted a tight system of cost control on every contract to restrict variations and extras that, if left unchecked, might have cost the university dearly. In all contracts there is a sum put aside for contingencies and it is very tempting for any architect to use this sum to refine the design rather than to keep it for unforeseeable emergencies, not necessarily informing the client beforehand. But, after some early misunderstandings had been resolved and mutual confidence established, Joe respected my responsibilities, kept me well informed, and worked with me willingly to keep within our budget. Although he spent countless hours devising designs for the undergraduate library, including the dozen that were rejected—all being meticulously detailed, and not sketch plans by any means, he astounded me by charging fees on the basis of only one design; that which had been finally approved.

I was devastated when he died prematurely, in May 1978 at the age of fifty-nine, before the university could consider awarding him a richly deserved honorary degree, and without learning about being elected as a Royal Academician shortly before his death. But Elain Harwood gives appropriately warm tributes to him and his colleagues in her sensitive work 'Chamberlin, Powell, and Bon'[2], in which she describes the Leeds Development Plan in great detail, considering it to be one of the firm's most outstanding achievements, and concluding with the words 'as a landmark in university planning it deserves to be better known'.

The university could not wish for higher accolades than those implied by Professor Whyte's paper[3], which makes complimentary observations about the university that I now recognise to be deserved but could never have imagined while I was immersed in the turmoil of events that are now regarded as historic.

Denys Lasdun had been 'runner-up' in our selection of an architect and was becoming recognised as one of Britain's leading designers. So, during a visit to London, I dropped by the new building that he had just completed for the Royal College of Physicians at the edge of Regents Park. It was startlingly elegant and would have blended with Chamberlin's work in Leeds to perfection, but I got no further than the spacious entrance hall, for, to my astonishment, behind the reception desk, sat the daughter of our Professor of Music. Jane Denny had spent some time with Margery and me, acting as assistant House Manager of Woodsley Hall, and was now the Secretary of the College of Physicians—a post for which her personality was far better suited. Such a small world.

The new Chairman of the Hospital Board, Sir Donald Kaberry MP was like a breath of fresh air. He had considerable influence in Whitehall, was keen to get on with planning the new hospital and medical school complex, and he promptly agreed to the joint appointment of the Building Design Partnership as the design team. The BDP were the first of our consultants to combine architectural design, structural engineering, mechanical and electrical services and quantity surveying in a single organisation under one roof, and the resulting efficiency was impressive. They are now an internationally famous company. Cham-

---

[2] Elain Harwood 'Chamberlin, Powell, and Bon – The Barbican and Beyond RIBA Publishing 2011
[3] William Whyte. 'The modernist moment at the University of Leeds 1957–1977' *Historical Journal Vol 51 2008*

Chapter 7: 1960–1978 The Chamberlin Plan

berlin had recommended them, and they were enthusiastic about his proposals for the location of the medical centre over the new ring road.

Unfortunately, the two clients did not achieve a matching state of harmonious efficiency, but through no fault of the Hospital Secretary or the University Planning Officer, for they both collaborated well. The inefficiency came from higher echelons, for the Leeds Hospital Board had to report to the Area Hospital Board, which in turn was responsible to the Regional Hospital Board, which in turn reported to the Ministry of Health, who provided the money. It seemed that *everybody* had to be consulted at *every* stage. An impressive joint planning team was appointed to work with the BDP design team. There were never more than four people on our university planning teams, usually Joe Chamberlin and Frank Woods on the one hand, and no more than two people from the Planning Office on the other. But the hospital's team was altogether more elaborate, with Sir Donald Kaberry in the chair, executives from the Area and Regional boards, the hospital secretary, matron and senior nurses, altogether amounting to fourteen people serving the hospital's interests. The university had two representatives—the Dean of Medicine and me.

The BDP usually sent their senior architect, Ed Hill, without whom the meetings could not function, of course. The same could not be said of the hospital representatives, who rarely opened their mouths, but patiently observed the proceedings. In practice, these monthly meetings merely served to rubber stamp decisions taken by executives who were in daily contact with the architects, but they took an inordinately long time, usually requiring a break for lunch. This made busy people like the Dean and me constantly fume about what was, for us, a complete waste of valuable time. So, when the Ministry of Health announced that planning operations were to be streamlined by management consultants, who would advise the hospital and the university on how to increase the efficiency of our working methods, we enthusiastically promised to cooperate.

At our very next meeting, Sir Donald, whose dynamic personality did not suffer fools gladly, could not resist making a caustic comment when the consultants arrived twenty minutes late for the first meeting thus wasting time, he said, of some twenty busy executives. They apologised and said that in future they would travel from London by air, which would be more efficient. Eyebrows were raised as to the extra cost, and they offered no response to the Chairman's comment that it would just give them extra time in bed. Their standing as experts in efficiency was further damaged the following month, when their plane was diverted to Teeside because of fog. When they did eventually arrive, halfway through the meeting and just in time for lunch, the Chairman, with a Herculean effort to constrain his language, made it plain that in future they must come up to Leeds by train the day before, spend the night in a hotel, and attend our meetings on time.

The consultants spend that afternoon giving an elaborate demonstration of a network diagram, which set out the stages involved in the planning process, linked by lines to indicate time periods. The objective was to uncover any sequence of operations that might threaten to delay the job, and to indicate any adjustments that could be made to smooth out wrinkles, to produce a Critical Path Analysis. It was convincing and beautiful in theory; just one more complicated job to be added to our, already heavy, schedule. We obediently produced an impressive network, which had a gratifying complexity, and which we hung on a wall to save desk space. But we found no practical use for it, and became so tired of wasting time explaining to visitors how the damn thing was supposed to work that eventually we hid it in a drawer. However, the hospital's network diagram did prove to be helpful in dealing with conflicting interests between consultants, nurses, ancillary staff, and the various authorities, in which its psychological leverage was used to stimulate people to make decisions.

I was lucky, for our professors were autonomous in their decision making, and I had no trouble in agreeing schedules of accommodation that were acceptable to them. The only contentious issue concerned the large departmental teaching laboratories, which threatened to absorb a disproportionate amount of the money available. They were occupied for only a few hours a week during the term and were empty for the rest of the year. The possibility of departments sharing laboratory space was a delicate issue, for prestige was involved; every department took pride in the size and sophistication of its very own teaching laboratory and they were not at all interested in the economics of a utilisation

Chapter 7: 1960–1978 The Chamberlin Plan

factor. So, I was delighted when the Dean of Medicine, Derek Wood, came up with a solution. He had heard that medical schools in North America, whose capital funding came from private sources, had developed a system of multi-discipline laboratories, shared by departments such as Bacteriology, Virology, Pharmacology, and Physiology, each of which had its own exclusive preparation room adjacent to the lab. Anatomy was the only department that could not participate in this shared arrangement, as they required highly specialised and immovable equipment, such as dissecting tables and so on.

Dean Wood had a quiet, unassuming manner, and was very well liked by the other professors, who considered him a great improvement on his predecessor (reputed to have once fallen asleep while he was chairing a committee), but even Derek's persuasive powers were stretched to the limit on this issue of sharing such important facilities. It was essential to reinforce his second-hand information with hard facts before the departments would commit themselves, and it was vitally important to get our design right the first time. So, it was arranged that he and I should tour American medical schools on a fact-finding mission before giving the architects any instructions. The Vice Chancellor wrote to the universities concerned, and the tour was all arranged when, at the last minute, the UGC[4] architect, Stanley Meyrick, asked if he might come with us. He realised that this concept could influence other UK medical schools, and he offered to help by expediting visas for the USA. When our passports were returned to us we were amused to find that we three now constituted a British Government Delegation, with imposing visas to prove it.

In Montreal and Toronto it was difficult to discern any particular difference attributable to our elevated status, but the University of Calgary did spoil us by holding a small reception in our honour. Here, in Canada of all places, I was amazed to see cars slithering about in some two inches of snow that had fallen while we were indoors. The drivers had evidently put their winterised vehicles and tyres away a bit too early. But, in Palo Alto, California , we found that spring had arrived. Our host professor said that he had intended to show us something on

---

[4] University Grants Committee

his computer, but unfortunately one of his research students was using it to write up his thesis. This was incomprehensible to me, for it was the only computer the department owned, and must have cost thousands of dollars. Couldn't the student have used a typewriter? This was my first intimation that computers could be used as word processors, but when Derek asked how the computer could be used for medical research, the professor just ignored his question and casually suggested that if we needed any refreshment we might use the student cafeteria.

Of all that we saw on this brief tour of North American universities, the campus that impressed me most was the one at Ann Arbor, Michigan, and I never imagined that I would come to have a grand-daughter Katie, who would graduate there. Nor did I dream that her sister would eventually live and work in Washington DC, where I received another culture shock. We alighted from a cab outside our hotel just in time to see a black sedan hurtling out of an adjoining side street and swerving on two wheels into the main road, followed by a screaming police car. I thought that this might be expected in a city like Chicago, surely not in the Nation's capital, but the cabby assured us that it was quite normal.

I managed to amass a vast quantity of information from which to select the best ideas from our North American experience, and the Dean was able to reassure his colleagues about the practicability of our proposals. However, Stanley Meyrick was no help, for he had relied on his routine of sending, in advance, a questionnaire for his hosts to fill in. Of course, this worked well enough in the UK where the universities depended on him for money, but the Americans told him bluntly that they had no time to fill out his forms. Some were more tactful, and implied that they had never received them. Stanley hadn't made any notes, and never did get around to making his own report, so he borrowed mine instead. He was evidently good at delegating, and I was informed that he was an expert in some obscure Slavonic/Russian language. He was later appointed Warden of Castle College, at Durham University.

Back in Leeds, Ed Hill and his colleagues at BDP were a delight to work with. They worked happily on the schedule of accommodation I prepared for multidiscipline laboratories on the basis of our tour. They also invited me to examine and discuss their first sketch plans

Chapter 7: 1960–1978 The Chamberlin Plan

for the Clinical Sciences block at St James Hospital, where the site was very restricted. I regarded this as a huge compliment, and they actually adopted a number of my suggestions. This was an example of the sort of admirable collaboration that BDP practiced throughout the whole of their extraordinary partnership.

I had become intimately involved in the design process a decade or so previously, when Allan Johnson and I were faced with the problem of our main boiler house. Boiler houses are not usually pretty; most are downright ugly, and as this one was to be in a prominent situation on Woodhouse Lane, the architect TA Lodge insisted that it should be put underground. But, as it was to be equivalent to three storeys high, I told him that this idea would be financially ruinous and ridiculously impractical. Lodge then became uncooperative and the design process ground to a halt. I was beginning to get very tired of Lodge; why on earth had we rewarded him with an honorary degree? Such procrastination might have resulted in the university losing the grant altogether, for the deadline was upon us. I put it to Allan that he and I should keep working in my office and not go home until we had achieved a sketch design to our mutual satisfaction. We both worked hard, and he got back to his hotel about midnight; I got home a great deal later. We were both pleased with our joint design, which I regarded as aesthetically satisfactory, although we could not avoid having a huge concrete chimney that tended to conflict with the august Parkinson tower, especially when viewed from the other side of Leeds.

The University Senate was sensitive to the fact that sulphurous gases were now replacing soot as the chief public health hazard to air quality, and went to the expense of buying expensive low sulphur oil from Canada to minimise the problem. So we were affronted to receive a notice from Leeds City Council to raise the height of the boiler house chimney from 120 feet to 150 feet! Evidently, the city medical officer had taken measurements and had concluded that our chimney was responsible for the excess acidity in our vicinity. We obtained estimates and submitted an application to raise the height accordingly, but to our astonishment and disbelief, the City Planning Committee refused permission on aesthetic grounds. We sat back and waited while politicians spent several weeks arguing about their differences, for prestige was involved and neither committee wanted to lose face. We were then told that the press were about to accuse the university of being irresponsible, and that this might result in legislation and bad publicity. At this, the Vice Chancellor saw red and demanded a joint meeting of the two committees, which he could attend, to expose their ludicrous behaviour. He persuaded them to compromise at a height of 140 feet.

It now appears, from a distance of some fifty years, that the BDP's magnificent new Central Electricity Generating Station (CEGS), originally intended to serve the new hospital and the medical school, is far more than adequate for these two buildings and is being used to provide heat for the whole of the university as well as the hospital. The boiler house of whose design I was so proud, became obsolete and is to be demolished; indeed the ugly chimneystack has already disappeared. So, it turns out that the aspirations of all the council committee members have been fulfilled, because the CEGS is gas-fired and pollution free. I wonder if any of the original committee members have realised this? I'm sure they would be pleased to know that all their arguments have been vindicated!

Chapter Eight

1948–1963: Conferences and Consultancies

In 1946, Mr Brown, the Bursar, introduced me to the University Finance Officers, whose conference was held in Manchester that year. Most of the delegates were accountants, who liked his suggestion that they might get rid of all their tiresome building problems, which were regarded as a damn nuisance, by delegating them to a resident Planning Officer, like me. Soon there were enough of us to justify our own Building Officers' Conference. Indeed, whenever a new university was proposed, a Buildings Officer was invariably one of the first appointments to be made. We met at a different university each year to borrow ideas and inspect new buildings, and I made many valued friends.

It was Professor Preston of Plant Sciences who alerted me to the dangers of asbestosis and prompted me to present a paper to the London Conference in the 1950s. The paper caused some consternation; it was the first time that members had learned of the serious problems of asbestos, then being used extensively for ceiling tiles, fire proofing, thermal insulation, sound insulation, acoustics, and so on. The post-war boom in construction worldwide had resulted in a phenomenal demand for asbestos, and mining firms in Canada, Africa, and Australia were making vast fortunes. They ruthlessly suppressed adverse publicity, even having enough influence to eradicate the word 'cancer' from an official medical report on the effect of asbestos on human health. They also refused to insert warning labels on their products 'in case it affected their sales'.

Since asbestos dust takes some time to manifest itself as a disease, doctors were slow to learn of its insidious effects, and then only among asbestos workers. By the 1950s it was becoming evident that this fragile material posed a threat to other people, not just the workers themselves. Even then, there was confusion and misinformation—when I wrote that 'blue' asbestos was the most dangerous, and implied that chrysotile and amosite were relatively harmless, I was mistaken. Not that it mattered; my colleagues got the message that all types must be treated as hazardous. Asbestos in any form is now banned altogether in the European Union, Australia, and New Zealand, but regrettably not in the USA, where profits can often take precedence over people's health.

Some fifty years after the publication of my paper, more than one thousand tons of asbestos dust was released when the twin towers collapsed on 9/11 and contributed to the high rate of cancer among rescue workers at the site.

Conferences were full of interest, especially when they were held in ancient colleges such as Oxford and Cambridge. At Durham, we learned all about their Bishops, their palaces, private armies, and foibles, and of one distinguished cleric whose portrait adorned the Dining Hall, known as 'Presence of Mind Smith'. Apparently, while boating on the river, his companion fell overboard and foolishly tried to clamber back over the side of the boat. As Smith proudly related to his colleagues, 'If I hadn't had the presence of mind to hit him on the head with my oar we might *both* have drowned!'

For years, our campus at Leeds was so chaotic that I put off holding the conference but when we finally did, it was a resounding success, because we had several buildings of which to be proud, particularly those by Chamberlin, Powell, and Bon.

Chamberlin's original plan had been so widely publicised that, when a more detailed report was published in 1963, there was widespread interest from universities all over the world. We became overwhelmed with visitors, who had to be conducted over our new buildings, and shown the Chamberlin Plan in the form of an illustrated lecture, followed by lunch in the senior common room. One day, a visitor from Canada, Professor Murdo MacKinnon, approached me with apologies for taking advantage of the free lunch in spite of missing the lecture. He wondered if he could make amends by inviting me to visit Guelph University in Canada where he was Dean; his colleagues and professional consultants would like to hear my talk. His university would pay my expenses, of course, and if I could bring my wife and daughter we would all be accommodated in the official guesthouse for a week. He assured me that he had generous hospitality funds intended for this sort of courtesy, and why not make a detour and take this opportunity to visit EXPO '67 that was being held in Montreal?

The Vice Chancellor liked the idea, suggesting that it would be a good opportunity to explore present thinking in the US regarding television as a teaching aid, and he promptly arranged interviews with professors at the Massachusetts Institute of Technology (MIT), as well as at Columbia University in New York and McGill University in Montreal. As it was a chance of a lifetime, I cashed an insurance policy to pay for our expenses, and Margery managed to book a room in Montreal, no mean feat, for the whole world was going to EXPO '67 and competing for accommodation at any price.

EXPO '67 itself was an ordeal, with long queues everywhere, especially for exhibitions that were worth seeing. As the humidity and heat of August were punishing, we skipped the less important exhibits and gave Sue a good time in the marvellous amusement park. As an eleven year old, she was quite willing to wait for twenty minutes to get on the 'Black Spider', which twirled and jerked us around until my knuckles became white to avoid being thrown out—an experience I vividly remember to this day. Our rented room had no air conditioning and was memorable for the delicious French cheese from which pungent aroma persisted long after we had consumed it. We remembered our meals in French restaurants long after we had forgotten the details of the exhibits.

At MIT, my interviews were enlightening; it was clear that the professors had very little respect for television as a teaching aid. Later on, when I saw a demonstration at the Rockefeller Centre in New York, I began to understand why they regarded television with such distaste, for the colours were garish and the images fuzzy, as one might expect from a technology in its infancy. Apart from this, they considered that operating a 'closed system' within MIT would be prohibitively expensive, and disproportionate to any benefits obtained. On the other hand, they considered that money invested in computer technology would reap rich rewards. They expressed reverence for the University of Manchester where Alan Turing, of wartime Bletchley fame, had built the first electronic computer. They were impressed when I told them that Leeds had one of the earliest computers, a KDF9, which we were busy linking by GPO coaxial cables (itself a dramatic new invention) to Manchester and other universities who also had KDF9s, to form a fully interactive multi-access system. In those days, a computer comprised rows of cabinets, requiring air-conditioning with a maintenance engineer on duty twenty-four hours a day! The demand for computing time was overwhelming, and the links with other universities spread the load and greatly increased their capacity.

The Americans' assessment of priorities was quite right; as I write this there are 9,000 personal computers and 256 'super computers' on the Leeds campus. Their sceptical view of television was embarrassing, but I need not have worried. After the appointment of Professor Blumler as the head of the Centre for Television Research in 1963, the idea of 'communications' involving photography, cinema, and television mushroomed at Leeds into a thriving Institute of Communication Studies. By 2010, this had achieved an international reputation and supported an academic staff of thirty, including seven professors with 700 students, including more than 100 postgraduates, and ten degree courses.

In Canada, Professor MacKinnon gave us a warm welcome, but his consultants seemed to resent the idea that 'the Brits' could show them anything about design. I had thought that the idea of covered links between buildings would appeal to them in the light of the severity of the Canadian winters. I was taken aback when they showed me the remarkable Scarborough College, near Toronto, designed by one of them, in which all the teaching rooms, study bedrooms and communal facilities were contained under one roof, in a building of striking and most attractive 'brutality'. It was very impressive, but did not have the flexibility that was built into, and was so important to, our much more extensive Chamberlin plan.

After the trip, I returned to Leeds and to the problems of re-organising accommodation for departments that evolved erratically. The

Mining Department had lots of spare space. It had originally trained engineers for the coal industry and the number of students was diminishing. No sooner had I suggested that some of their surplus accommodation might be used by other departments, who were badly in need of space, the department renamed itself the 'Department of Mining and Mineral Sciences' and proceeded to order lots of new equipment, and started agitating for an extension to be built.

This behaviour was not unique. Other ideas of mine, such as a re-allocation of surplus space in the old Agriculture building had to be abandoned when the degree course in 'Agriculture' was replaced by four new courses; 'Agricultural Science', 'Agricultural Botany', 'Agricultural Zoology', and 'Agricultural Biology', with appropriate increases in the number of staff and students and of course, demands for more rooms.

Despite these setbacks, I really did hope to acquire some space in the ancient Department of Leather building, designed in the 1880s, when there was a prosperous industry manufacturing boots, saddles, luggage, and leather goods of every description. Professor Atkin, a charming man renowned for his repartee at lunch in the old refectory, had only a few research students and a small group of undergraduates on a course that had not changed for many years, so the closure of his department was being seriously discussed. However, as it remained the only department of its sort left in any British University, the Guild of Leatherworkers was fighting fiercely to preserve it.

My hopes were completely dashed when some genius came up with the idea that biophysical aspects of leather related so closely to food manufacture that it justified developing a course on 'Food Science'. So, on Atkin's retirement, a dynamic new professor, Donald Burton, was appointed. He soon had things humming, not only scotching my plans for acquiring space for other departments, but appointing enough additional staff to justify expanding Food Science into another building that had just been vacated by the Electrical Engineers. This was three times the size of the old Leather Department building, and before long the department of Food Science filled *both* buildings and had achieved an international reputation for research, especially in the field of frozen food.

One day, the Vice Chancellor called me up to his room, referred to the success of our new refectory, and asked 'would I collaborate with Food Science in advising the University of East Africa on their catering arrangements?' I was dumbfounded. I pointed out that my office was already overwhelmed by the implementation of Chamberlin's plan, and with the management of contracts we had with other architects. But the Vice Chancellor said that he had already committed Leeds to provide a consultancy; 'our prestige is at stake, and you need by away for no more than two, maybe three weeks'. So it was settled! 'How quickly could you arrange to go, for this is urgent?' (I had yet to be faced with a project that wasn't).

It transpired that our Government had agreed with Denmark and the United States to help with the expansion of the University of East Africa, and, in particular, with the University College in Nairobi. The Danes were to provide a development plan, the Americans would provide student housing, and the British were to be responsible for catering, in the light of a recent report by Woods and Tower recommending the creation of a central catering unit, to replace all the kitchens in the existing refectory and halls of residence.

The food scientist appointed to work with me was George Glew, who was running a catering experiment in the small Women's Hospital as a pilot scheme for our new Teaching Hospital in Leeds. This was similar in principle to the original proposals for the Southern Hospital in Stockholm, but was far more sophisticated, in that prepared meals arrived outside each ward in a frozen state, to be reheated in a small server and then served piping hot at the bedside. George took me for a meal in the Women's Hospital; it was delicious, for not only were the best ingredients used, the cooking was carefully supervised, and the timing of the re-heating ovens was automatic, leaving nothing to chance.

George explained that the journey from a central kitchen to the patient's bedside was only one facet of the whole problem of catering in large hospitals. Quality control was difficult to maintain when large quantities of fresh meat, fish, and vegetables were arriving for immediate conversion into hot meals—perhaps 2,000 or more each day. A hospital chef would think twice about rejecting any raw food that was marginally

Chapter 8: 1948–1963 Conferences and Consultancies

acceptable in quality, for any dislocation of the cooking process would be the very last thing he wanted. Large kitchens began with a relatively slow start in the mornings, with a gradual acceleration of effort, culminating in a frenzy of activity just before and during meal periods.

With George's frozen food system, the cooks would systematically prepare meals, freeze them, and put them straight into storage, with no pressure to meet a critical deadline. Any deliveries of meat or fish considered substandard could be rejected and fresh supplies ordered. In the meantime, other meals already cooked and deep-frozen could be brought out for that day's menu. The cooks could work smoothly through an eight hour shift without any distractions or panic. The number of kitchen staff could be reduced in view of the even tenor of the work, and sick leave and holidays need no longer result in paroxysms of anxiety in the kitchen. This way, a larger variety of meals could be kept available in storage, including those for patients on special diets.

I now realised why the Vice Chancellor was keen to publicise this breakthrough in technology, and why the Ministry of Overseas Development was prepared to finance such a prestigious pilot scheme abroad.

As George was tied up with teaching and research for the next few weeks, I had to 'go it alone'. Guided by Margery to a specialist outfitter in Jermyn Street, (how did she come to know of such a place?), I bought a 'bush hat' to guard against the equatorial sun and left on a foggy November day in London, to arrive somewhat guiltily in the warm sunshine of Kenya, with fleecy cotton clouds and a gentle breeze. No wonder the white settlers called it 'Gods own country'—no winter, no summer, at least three crops a year and a blissfully pleasant temperature day and night.

The Principal of the College gave me a warm welcome, and a friendly catering manager promised to help all she could, but first there were the obligatory courtesy meetings with the British High Commissioner, important academics, and a formal evening dinner at the Principal's official residence, where the chief guest was the Ghanayan Ambassador. At first I was self-conscious, being the only white person present, and unable to make out a lot of what was said, but I needn't have worried, for there was no formal discussion at all; everyone was intent on enjoying a splendid meal and having a good gossip.

The Ambassador's wife set the tone by recounting her first experience of polite society. At that formal dinner, when the hostess invited the ladies to leave the gentlemen to their port and cigars, she decided to remain with the men, for she was rather partial to a glass of port, and never refused a cigar. To her annoyance, her hostess insisted that she should come upstairs with the other ladies to 'powder their noses'. When she happened to glance through a window overlooking the garden, she saw in the moonlight seven streams of silvery liquid descending on the rose bed as the men relieved themselves, and realised why she had been invited upstairs. The guests at our dinner were convulsed with laughter, especially when, in reply to a question, she said, 'Yes, I always go upstairs now—you get a much better view up there!'

It was a jolly evening, and I wish I could have understood more of their infectious humour, but I had a lot of work to do, and left as early as I decently could. The Principal insisted that I should be driven the half-mile or so to my hotel in his limousine because it was 'far too dangerous to walk there alone'.

The College had booked me into the famous English Garden Hotel, where I was beginning to feel very comfortable until I was visited by an old colleague. Royston Jones had been a lecturer at Leeds and was now Professor of Civil Engineering here. His wife and daughter were in England, and he begged me to stay at his house to keep him company. Although his home cooking was a poor substitute for the hotel's fare, I have to admit that his knowledge of local affairs and guidance about the city was invaluable in my work, and I was grateful for his hospitality, especially when he introduced me to his good friend, Guyron Aston.

Guyron was advising the Kenyans how to organise their local government; they liked the electoral system that distributed privileges more evenly than before, but they had no sense of communal responsibility. He described how one city councillor, immediately after being elected to be Chairman of the Works Committee, walked up to a repair gang in the street, told the bulldozer driver to follow him, and on reaching his own house instructed him to make a road joining his property with the nearest public street. Guyron knew that by tradition, this process would have continued until all the members of the chairman's family had new roadways to their houses, and had to explain that this was not appropriate behaviour. With every facet of local government reflecting this traditional approach, I could see that Guyron had his hands full.

Both he and Royston expressed concern when I admitted to having surveyed the whole of the university campus on my own; it was a large area of parkland in the centre of Nairobi, and they thought I'd been very lucky to escape mugging. They went on to tell me about petty pilfering, towards which there was a generally benevolent attitude, because the Kenyan economy was in an appalling state and most of the unemployed in Nairobi kept their families alive by thievery. Some had developed a sophisticated system using fishing rods to poke through the tiniest crevice, hooking valuables with great skill, so windows in houses and cars had to be kept tightly shut. The police were not interested in solving such petty crimes; they preferred to patrol the main highway between the city and the airport, where unsuspecting drivers could be apprehended for a variety of infringements of the law and a cash fine extracted on the spot.

Each morning I woke to hear the men working on a hostel being built next door. They were so pleased to have a job that they started hammering well before the official starting time, but it was not so much the noise of hammering that woke me—it was their joyful singing in harmony that delighted me, and which continued on and off for most of the day.

My commission was to prepare a brief for the central catering unit, so the first job was to examine the scale of the problem and to suggest a location for a central kitchen within the Development Plan. However, when I asked to have a look at the Development Plan, I was told that there wasn't one; the Danish Government hadn't even got around to appointing an architect to create one, and although the Americans had undertaken to build more student residences, no one could tell me what sort and size they would be or where they would be located.

The only thing I could do in these ridiculous circumstances was firstly concoct from statistics an estimate of future student numbers, secondly, make a guess at what some Danish architect might suggest for a master plan, thirdly, imagine what the Americans were likely to come up with, and fourthly, propose a site for our building, which would not seem stupid when these people came along with their ideas.

When I discussed this ludicrous situation with the Principal he was sympathetic, but hoped that I would produce a report that at least would start things moving. I had no option but to put in long hours of hard work in view of the short time allotted for my visit.

As the second weekend arrived, Royston insisted that I must have a break, and Guyron took us down to the Tsavo Game reserve in his tough old Mercedes. The car bounced around on the deep ruts and potholes in the murram roads until he accelerated to 70 mph, when the ride became unbelievably smooth—so long as one hung onto the roof handles!

Tsavo is renowned for its lions, and Guyron did them justice by having the biggest camera I have ever seen, taking 70 mm transparencies (similar to movie cameras) through a gigantic telephoto lens mounted on a bracket on the side of his car. He already had a superb collection of slides, but he wanted even better ones, and so we pressed further into the reserve looking for suitable prey for lions (impala or gazelle) or circling vultures that would indicate a kill. With my little 35 mm telephoto lens I shot giraffes, wildebeest, waterbucks, monkeys, and crested crane flamingos until I ran out of film.

Guyron slowed to a crawl when we found ourselves trailing a huge bull elephant. He warned us that such a single elephant was dangerous; we must be ready to reverse and 'get the hell out of it'. As we were on a narrow track with tall grass on either side of us, I didn't like the sound of this, but he kept a safe distance until he could alter our route. Just as dusk was falling we spotted a solitary lioness silhouetted on a slight rise. We kept to leeward and saw that she was keeping an eye on a herd of antelope, with the rest of the pride out of sight.

The next day, we saw hippos in the river, and Guyron explained that they came on dry land to graze, usually at night, but returned to the river to take the enormous weight off their legs. Their excreta provided vital nutrients for the fish in the river, but it had to be prepared for them and was deposited with surprising sophistication. The hippo has a short, powerful tail that, when wagged, sprayed the excreta over a wide area. As Guyron said—'it's not the usual case of the 'shit hitting the fan', but of 'the fan hitting the shit!'

We saw more wildlife than I had ever imagined, including a pair of rhinos that Guyron kept at a very safe distance. He had once been

Chapter 8: 1948–1963 Conferences and Consultancies

We came across a huge bull elephant in Tsavo National Park.

charged by one, and always approached rhinos by driving backwards, in reverse gear, ready for a quick getaway.

After fifteen days' intensive work I presented a twenty-seven page report, with maps, tables, and diagrams, to the Principal and his colleagues, outlining student numbers, describing frozen food operations, and making recommendations for future action. I was describing the final page when I keeled over and fell on the floor. I was found to have a high temperature and had visions of some horrible tropical disease—but it was only dehydration and the following day I flew home. As I went through Customs, I found myself trembling and sweating again, which made the Customs' officer very suspicious, but his meticulous examination only revealed a parcel of Christmas presents. Royston had asked me to post them for him when I arrived home, for when it came to handling parcels, the Nairobi Post Office was not renowned for honesty.

George Glew eventually visited Kenya, and the Principal came to Leeds with his wife, saw the Food Science department and sampled a frozen meal. He approved the sketch plans for the central catering unit, which we were developing in collaboration with a London firm of architects, and Alan Lolley, of my office, spent many hours preparing schedules and estimates of all the special blast freezers and reheating ovens required in the kitchen and serveries.

It was intended that once the building was ready for occupation, George should go out and supervise the start of the whole procedure with me in tow, but when we asked the architects about the progress of the work they were strangely reticent, and it eventually turned out that the College had abandoned our idea of frozen food and had instructed the architects to design the refectory with traditional facilities.

I must say that, on reflection, the College had taken the right decision. The 'men from the ministry' were misguided in their efforts to sponsor this particular British expertise. For whereas the frozen food concept proved to be perfect for hospitals in Britain, where labour costs are high and electricity is relatively cheap, the opposite was true in Kenya. I recalled the catering manageress telling me how the students, coming from remote villages where they were treated like royalty, soon developed exaggerated ideas of their own importance. Enjoying four free meals a day (including afternoon tea) and learning how to use table cutlery, they often demanded more food than they could eat, and either left the surplus on the table or on the floor. As the cook-freeze system was based entirely upon uniform meals of a regulated size it did not make cultural sense.

If only George and I could have gone out together with time to examine the problem properly, by talking to all the people involved, we might have saved the Government from an embarrassing diplomatic blunder, which of course nobody mentioned. However, my directive had been unambiguous—I did exactly as instructed—it had been a mad rush for me from start to finish, and my pathetically inappropriate report now languishes at the very bottom of my bottom-most drawer.

The fame of Joe's development plan spread beyond the Iron Curtain, and I was invited to go to East Germany to meet the architect responsible for buildings at the University of Leipzig. Ron Crawford volunteered to come with me. At that time, the German Democratic Republic (GDR) had a formidable reputation for shooting people trying to escape over the Berlin Wall. Even flying over that wall was a bit unnerving—we found the very drab East Berlin airport building occupied mainly by military personnel and large men in grey overcoats who seemed to have nothing else to do but stare at us with suspicion. We wasted no time in going through customs and passport control, and were relieved when a young man came forward to greet us. Hans was a junior member of the Department of English and was overjoyed to practice his language skills on the first English people he had ever met.

He explained that he had the privilege of acting as our guide and interpreter for the whole of our brief stay, and to ensure that we enjoyed our visit and complied with the regulations. If we did not, he would get into trouble and above all, we were not to go anywhere without him. He was such a gentle creature, who looked as though he needed a full English breakfast to get rid of his pallor, that we had no hesitation in promising to be compliant.

The drive from Berlin to Leipzig was particularly interesting to Ron Crawford, who had been a bomber pilot during the war. We soon realised what a mess East Germany was still in, even though the war was long since over. The only manpower available for clearing up and rebuilding consisted of schoolboys and men too old for military service. Thousands of East German soldiers were still being held as forced labour to repair the damage suffered by Russian cities. Most of those who surrendered to the Allies had sought asylum in the West, and very few had returned to Leipzig. Consequently, there was a desperate shortage of labour, and the only way the university could be rebuilt was by making it a condition that all students must spend at least one year working on one of the university building sites.

Hans was particularly proud of being able to take us to our hotel in one of the university cars, the ubiquitous Trabant; the only car being manufactured in the GDR and the only make we ever saw. The East Germans had no foreign currency with which to import cars, and the public had to wait several years to buy a Trabant. They were treasured, and with careful maintenance, their average life span was twenty-eight years. They did not have a fuel pump; the gas tank had to be sited higher than the engine so that the feed was by gravity, and there was no petrol gauge, so a dipstick had to be used to measure the amount of fuel left in the tank. The two-cylinder engine seemed to perform well enough on our short journeys, but in the very bitter weather it had a job to create any warmth. This gave Hans the excuse to leave it running, even when he might have switched it off, admitting to us the real reason; he was afraid to stop the engine, in case he couldn't start it again!

The Trabant was the only car being manufactured in the GDR and the only make we ever saw.

After a very bumpy ride along roads ridged with frozen snow, Hans escorted us through the hotel's reception procedures, where our credentials were scrutinised by two very large men wearing overcoats and trilby hats, although it was quite warm in the lobby. Here we were relieved of our precious passports, and when Hans noticed our perturbation he assured us that they would be returned to us eventually. When we were established in our room, he asked us in an unnecessarily loud voice to confirm that we would not leave the hotel without him. For the benefit of any microphones, we reassured him. He declined our invitation to dine with us, and on the way downstairs, he explained that room service was available from the attendant who was stationed at the end of the corridor. As this formidable lady was built more like an all-in wrestler than a waitress, and as our German was very limited, we didn't make use of her services. Shortly afterwards we found ourselves unaccountably speaking in whispers.

The next day, Hans appeared at our bedroom door at 8.00 am and introduced us to Herr Helmut Ullmann, who was one of the chief architects in the East German Ministry of Reconstruction and responsible for the new university buildings in Leipzig. These were extensive, and he showed us several ingenious stratagems that had been devised to compensate for the shortages of steel, cement, and fuel. Happily, the architectural style of his designs bore no resemblance to the stodgy buildings we had come to associate with communist regimes, but I have no pictures, for cameras were forbidden.

After lunch, he took us over the Leipzig Opera House, gutted during the war and restored under his direction. It had reopened in 1960. It was impressive, and a tribute to the aspirations of a people who had given priority to its restoration, while struggling to make their domestic accommodation tolerable. The East Germans were also fanatically devoted to success at the Olympic games, and sports fields and gymnasia were a high priority. Any student showing athletic promise was excused from work on the university building sites, and spent that year developing his or her physique and technique (1968 was the first year in which the GDR was allowed to participate, and they came fifth in the medal-winning stakes).

After years of suffering the results of war with Hitler, neither Ron nor I had any love for the German people as a whole and the current communist regime with its secret police did not improve our outlook, but Herr Ullmann and Hans proved to be two very humane people who were trying to make things better in spite of all the difficulties they faced. We had no hesitation in inviting them back to visit Leeds if they could get appropriate visas, and, in due course, this was arranged.

I never knew why, but their visit to Leeds was kept low key, and apart from the traditional guest lunch in the Senior Common Room, the guests were left in the hands of Ron and myself. This suited Herr Ullmann very well, for his English was worse than my feeble German, and we both became entirely dependent on Hans who was rapidly becoming fluent and indispensable. Herr Ullmann was suitably impressed by our buildings, and especially by Chamberlin's Development Plan, but both he and Hans were surprised and pleased when I took them to Summerbridge, where Margery not only provided excellent meals, but put them up for the night; they had expected to be fobbed off with one of the hotels in Leeds. I shall never forget the delight with which Hans greeted us at breakfast the following morning. He had risen at some unearthly hour and had gone exploring into the village, going into the shops and meeting people on the street. It was the very first time in his life that he had been out of East Germany and able to talk freely to anyone he liked. He was quite intoxicated with the experience, and it soon became clear that both our guests would like nothing better than to be shown something of our English heritage. So, off we went on our favourite route, via Fountains Abbey, Ripon, and up Wensleydale to the quaint old church at Wensley with its magnificent curtained pew for Lord and Lady Bolton. There were exclusive pews, at a lower level of course, for their family and servants too. Then it was on to Middleham Castle, which King Richard III regarded as his home.

The first time I had taken this route was just after VE day, when Father had booked a week's holiday at a guesthouse in Askrigg, where we arranged to meet. It was a Saturday, and the overcrowded train had stopped at every station in the two hundred mile journey from London. Another train journey and two bus rides took me to Leyburn by nine at night. I enquired at the police station, but there was no way to get further up the dale, so I telephoned Father. He told me not to worry; I was to be picked up by the owner of the hotel in Bainbridge as soon as she closed the bar. She finally turned up at about midnight, sweeping through the market square with considerable panache and screeching to a halt in front of me—I was the only person in sight. Father emerged from the car somewhat shakily. He was as white as a sheet, and whispered that our chauffeur had been drinking along with her customers at the bar, and he hoped that we would arrive safely back at the hotel. I soon found out what he meant. The road up the valley rose and fell, twisted and turned enough to test an experienced rally driver, and Father and I were thrown about in the back seat to the noisy accompaniment of several empty bottles that kept rolling over our feet. There were no safety belts in those days and nothing to hang on to, but our hostess was a cheerful raconteur who regaled us with a series of anecdotes as she drove the car at breakneck speed, explaining that she used the route so often that she knew every inch of the road and every bump and pot hole—of which there were plenty.

She certainly was a superb driver when drunk, and I wondered how she would perform when sober! To this day we still owe her—she refused all our attempts at reimbursement, so long as we promised to

Chapter 8: 1948–1963 Conferences and Consultancies

stay at her inn at Bainbridge the next time we were in Wensleydale. However, every time we tried to do so she was fully booked, for she was one of the outstanding personalities in the dale and her inn was very popular.

It was during that holiday, while taking a walk around Askrigg, that I came across Nappa Hall, a fortified farmhouse whose owner, Mr Metcalf, told me that his family had lived there for centuries. Much later, I found a record that in the fifteenth century the Metcalfs had attended banquets held by King Richard III at Middleham Castle, along with the Scrope family from nearby Castle Bolton (from which Mary Queen of Scots had been confined after her escape from Scotland, and from which she had escaped, making it as far as Leyburn before being recaptured. The hill named 'Leyburn Shawl' is so called because she dropped her scarf there). King Richard hated London and the Royal Court, and was always keen to get back to his beloved Middleham, where he had been brought up and educated. Recently, in 2014, King Richard's body was discovered buried under a car park in Leicester and re-interred, with due ceremony, in Leicester Cathedral. Arguments about his true character continue today—I am sure he was not as bad as Shakespeare painted him.

From Middleham, Ron and I, together with Herr Ullmann and Hans, went 'over the tops' and down to the medieval Skipton Castle, built in 1090 by Robert de Romille, a Norman baron. We arrived just as the owner of the castle, Mr Fattorini, was closing down his shop and cash register. My dismay must have been evident, for when I introduced our distinguished foreign visitors, who had to begin the journey back to East Germany the following day, he immediately welcomed them in. He gave us a tour of the whole castle, including staterooms, which he occupied with his own family and were not open to the public.

The next day, Ron and I decided to take our visitors down to Heathrow via a scenic route though the Wye Valley, stopping at Stokesay Castle and going on through Ludlow to Bath, where they decided they had had enough of British history and were delighted to go off on their own to do some gift shopping. For Herr Ullmann, the brief interlude in Bath proved to be the highlight of his visit, but not on account of the venerable history of the city or its architectural treasures. He came

The owner of Skipton Castle gave us a personal tour, which included his family's personal staterooms.

to dinner bubbling with excitement, having found a 'Do-it-yourself' store—a sort of shop that he never imagined existed! He was determined to insist that his government open one in every town in the GDR. The ability to buy small quantities of timber, plywood, nails, and screws off the shelf would allow householders to get on with small jobs themselves instead of having to wait for weeks on end for tradesmen to turn up as at present. The idea of renting small power tools for a day or two was just the icing on the cake. He said it would revolutionise life for the ordinary person—he could not wait to get back to Leipzig fast enough. Herr Ullmann had a home in the countryside and invited Margery and me to visit him anytime. The prospect of another few days behind the Iron Curtain did not appeal to either of us though, and we were content enough to think that he had enjoyed his fruitful time in the West.

Chapter Nine

1962–1992: Woolwich Farm

When we found the farmhouse for sale in Summerbridge, as described at the end of chapter six, an old man had advised us not to go up the cart track to see the old place. We ignored his advice. At first sight, the farm was depressing—dilapidated and unkempt, with a wooden garage in a state of near-collapse, a tiny greenhouse with its glass broken, and a field littered with hen huts. But, the quality of the farmhouse took my breath away with its dressed stone, massive quoins, stone roof slates and, above all, mullioned windows. Even the exterior earth closet (which housed bats) and the outbuildings had been built with monumental masonry.

In spite of its venerable age (built around 1725), there was no sign of subsidence in the walls. Peeling paint and rotten window frames were trifles that were not going to deter us; it was just what we wanted. Our elation was punctured when the estate agent told us that it was already sold. 'Had a contract been signed?' 'Err, no' he admitted. 'Has a deposit been paid?' 'No, there was some haggling about the price'. 'Right', I said, 'here is a cheque for the deposit—when can we take a look inside?'

Inside there was beetle infestation in all the beams and rafters but the roof was sound. My parents would have said that we were quite mad, but we reckoned that the kitchen, living room, toilet, and dairy, with two bedrooms and a bathroom upstairs, were just about adequate for us if we didn't have any visitors. It was perfect, with complete privacy and magnificent views over the upper Nidd valley, for we were surrounded by fields and woods in every direction; the only houses in sight being far away on the other side of the dale. When we examined the deeds we became really excited. I had always dreamed of owning a field with a stream in it, and here we would have two fields, two streams and two woods, with our own fresh water supply from a spring, and a pond with moorhens nesting on a tiny island in the middle of it! So we clinched the deal.

The old man had been right about the deeply rutted cart track, impassable for our mini, which had a clearance of only about twelve inches. We bought a Land Rover, but it was too hard sprung for Margery's back to tolerate and it was quickly returned. There was no option but to re-build the whole half-mile length of the track so that a normal vehicle could navigate it, and then we could move in. Laurie Gill, the farmer who owned the fields the track passed over, was co-operative and offered to pay for a cattle grid at the bottom of the hill. He must have foreseen my inability to get a car through the gate without letting his cattle escape on to the main road!

Our old home, Pasture Cottage, was advertised for sale, but then we were rudely shaken by a body-blow. The farm immediately adjoining Woolwich was put up for sale, and the potential buyer applied for Planning Permission to use it as a caravan holiday site. Owen Well

We were thrilled at our first sight of Woolwich Farm, although it presented serious challenges.

was a charming old farmhouse, probably older than Woolwich, and its thirty-five acres stretched right up to our front gate. We could imagine caravans in the adjoining field, and dozens of children exploring our woods and fields. What would happen to our privacy and tranquillity then? Quite horrified, we set about collecting signatures to support a protest at the upcoming Public Inquiry.

We knew that not many people would be as directly affected as us, so it was a relief to find that the Barretts of Braisty Woods farm, just on the other side of Owen Well, were also strongly opposed to the idea of a caravan site. No one knew *us*—we didn't even live in the village, but during our desperate canvassing we met lots of nice, sympathetic people, and collected an impressive number of signatures. It was a nerve-wracking ordeal, and when the inspector decided in our favour, our relief was indescribable. We had committed every penny to buy Woolwich, and many more anxious months were to pass before we were able to sell Pasture Cottage and get some money back into the bank.

Once the lane was usable by car, and a 'party line' telephone installed (phones were rationed at that time, and we had to share the line with Owen Well), we moved in. There was so much to do, and money was a real problem. The Government, faced with post-war reconstruction, did not allow banks or building societies to grant mortgages or loans for properties over fifty years old. This meant that all the money needed for improvement and repairs had to come from whatever money we could scrape together at the time. The first priority had been the lane, for motor access. The second priority became self-evident when the living-room carpet started to go mouldy! The previous owners had overlaid the original earth floor with concrete, on the assumption that this would stop any rising damp. What pathetic optimism! The water seeped through the porous concrete, and crept up the stone walls that had no damp-proof course in them. Traditional loose rubble infilling made the insertion of a damp-proof course impossible. I decided to lay thick plastic sheeting over the floor and up the lower part of the walls, and a layer of concrete to hold it down, with electric cables embedded in the concrete to form under-floor heating—a new invention just then being pioneered by the Yorkshire Electricity Board.

Our first experimental floor in the kitchen was such a resounding success that over the next several years we used the method in most of the ground floor and, much later, in the dairy and front porch. Electric under-floor heating seemed luxurious, but we found it to be very economic when used in conjunction with wood burning stoves, and the dogs loved it. Bare concrete floors in an eighteenth century farmhouse would have been an abomination; we had to have York stone flags, such as those in the dairy. Enquiries led us to Barry Foxton, a local builder, who became a stalwart friend. He was enthusiastic about restoring Dales properties in a traditional manner, and undertook to carry out work as and when we could afford it. He remarked that new stone flags were expensive, and he could supply old ones worn to a lovely patina by years of wear, far superior to new ones, and cheaper. They were perfect for the electric under-floor heating system, and looked superb in the main entrance hall and corridor.

The daylight from the mullioned windows was quite inadequate for the whole kitchen, so we asked Barry to break through the south wall to insert a new window. When he started the work he was astonished to find a large bread oven with a domed roof (a 'beehive' oven) contained within the wall, which itself proved to be more than three feet thick. He ventured further and revealed an enormous open fireplace measuring about seven feet across. This was easily large enough for us to install a huge Scandinavian wood-burning stove, with space on each side where one could look up the chimney and see the sky. The beehive oven had its door at the right side of the fireplace, and we used it for storing logs. There was also a little recess on the left, presumably for the dry storage of that once scarce and expensive item—salt. When we first moved in, this wall had been completely covered with plaster and sported an ugly tile fireplace from the 1930s!

The transformation of what had originally been the sitting room, turned by us into a kitchen, and now revealed as the large original kitchen was astounding. Barry managed to insert a small window and fan above the beehive oven without damaging its domed roof, which now formed a most attractive architectural feature. For years afterwards I always got a thrill when walking into that room. It had been restored to what it had always been—the focal point of the whole house. Indeed, we enjoyed the rough stone so much that we decided to remove the plaster from the corridor too and were amazed to find that the plaster was original eighteenth century horsehair plaster, evidenced from the tufts of horsehair sticking to the stone itself!

The major problem still to be tackled was rising damp in the walls, and Rentokil, who had dealt with our death-watch beetle infestation, told me of their latest invention—electro-osmosis. This consisted of thick copper strips embedded in the wall, and grounded by copper rods drilled deep into the ground. It was an expensive and noisy process, and people on the other side of the dale complained about the incessant drilling, but when finished, it worked perfectly.

It was more than three years before we finished all this building work, and could start raising funds to make a third bedroom upstairs. This was a major undertaking, for the only available space was in the hayloft, which had mullioned windows and a magnificent oak truss supporting the very heavy stone roof. The unsafe floor needed new joists and floorboards, and we enlarged the windows from two-lights to three, using old mullions that Barry had found. He was now showing a proprietary interest in our house, scouring the district for genuine antique components. He always had a skilled labourer to help, but did all the important work himself with meticulous care, and was as pleased and proud as we were with his achievements.

To reach the hayloft, we made a corridor through our bedroom, and cut an opening to make an open balcony overlooking the staircase and entrance hall below. To our delight, the underlying structure was most compliant, exposing huge timbers that seemed to be expecting this sort of treatment, and the result gave a mediaeval feeling as one entered the entrance hall below, especially when we later exposed the oak rafters in the ceiling. The resulting new bedroom was unexpectedly magnificent—far too good for visitors—so it became our bedroom for many years until I retired, when I used it as my studio. By that time, after twenty years of continuous effort, we had got round to adapting the old bothy beneath it into a spacious sitting room with a wood-burning stove. We were beginning to realise the full potential of this lovely old property, but there was one major snag; we could only get into our nice new sitting room by going outside, through the garden. This was a nuisance in winter!

We put up with this arrangement until we could afford to break through two thick walls to gain access from the old dairy behind; a cold, damp room housing a huge water tank, filtration plant, and a noisy old water pump, plus a clothes washer, spin drier, deep freezer, and other utilities. The stone shelves were always stacked with an unsightly assortment of bottles and jars storing our home made jams, pickles and wines, a pail of eggs being kept in water glass, and the walls usually had carcasses of pheasants and hares being hung before cooking. Guests being escorted through this 'foyer' to the sitting room found it eccentrically entertaining, but we felt embarrassed by the noise and permanent untidiness.

So, there was no alternative but to shift the water tank, pump and other equipment to an old stable, north of the bothy, which involved diverting the water supply pipes and hiding them in the walls. What, then, to do with the now-empty dairy with its leaking roof, damp floor, and the surrounding earth reaching half way up the walls designed to keep it cool? Damp proofing would be no small matter. However, by using my lump sum retirement pension, Barry laid a stone-flagged under-heated floor, built block walls to create a cavity for the pipes, constructed a new roof, and installed new windows, taking care to retain one of the original stone shelves that had been a feature of the old dairy. To our surprise and some disbelief at the way things had turned out, this became our favourite room, a comfortable, cosy little snug, especially in winter. It was then, after thirty years of strenuous and relentless effort, we felt able to relax, for nothing else needed to be done to the old house now that it was no longer too disreputable for visitors to see, and was a delight in which to live.

Of course the three wood-burning stoves had to be fed, but we had over three acres of ancient woodland with plenty of dead wood and fallen trees to harvest. For the first two years, I employed the local forester, Mr Lambert, and his son to climb the big trees and trim the highest dead branches. After the acquisition of the tractor I tackled the felling of dead trees myself and logged them and split them for burning. It was foolhardy of me to tackle this sort of thing without helmet or means of calling for help, and over some twenty years I had some narrow escapes that I shall never reveal, but it was extraordinarily satisfying work. Imagine my hilarity when I discovered much later on that our house in Illinois was next to a road named 'Lambert Tree'.

Over those thirty years we had experienced the joys of living in the country. On the first day after we moved in, Margery had disappeared with her two black spaniels, Brig and Jenny, I had assumed for a walk in the countryside. She returned radiant with excitement, saying she hadn't gone far, only a little way into our own wood. She had found a fallen tree in a small clearing on which she had been sitting in complete privacy, happily listening to the

A passage of flagstones led to our garden porch.

silence, broken only by the hum of insects and the songs of the birds. That clearing became her very own retreat, symptomatic of the peace and tranquillity we found at Woolwich, and a foretaste of the joy we would have from our association with all the birds and animals that shared our land. Indeed, at the end of our life there, one of the oaks near her glade was identified as the oldest in Nidderdale—over 800 years old!

Our original canvassing of the neighbourhood, even before we had moved in, had introduced us to many local people who became friends, and we were taken aback, some twenty years later, when a total stranger in the village asked 'Are you the folks who stopped the caravan site?' Fortunately for us, Owen Well had eventually been bought as a weekend cottage by Henry and Nora Hartley, a charming couple so modest no one would have guessed they were owners of a huge mill in Bramley, near Halifax. They leased their land to Arthur and Marion Barrett of Braisty Woods, whose farm was the largest in Summerbridge.

It was Arthur, a gentle and quiet man, who politely pointed out after a couple of years that it might have been easier and cheaper for us to have improved the lane from his farm at Braisty to Woolwich, which was on the level, rather than the one that climbed the hill up from the village, that we had chosen to repair. Each was about the same length, about half a mile long. When we had first set eyes on it, ankle deep in mud and almost impassable for cars, it had appeared to be useless. I explained that we were not in that class of affluent city dwellers who were buying up country cottages for use at weekends, and that we had no money available to rectify my error of judgment in the matter of roads. We were scraping money together to restore Woolwich to live there permanently.

Arthur said he would do everything he could to help us, and the Barrett family proved to be the most wonderful friends and neighbours one could wish for. In fact there was only one section of the lane that was really impassable. A picturesque waterfall was pouring over the wall from our wood, creating a permanent quagmire in the lane. The stream had originally passed under this wall into a drain below the roadway and into the field beyond. All the drains had been completely clogged up for ages, and soil had steadily built up behind the wall until the stream reached the top and cascaded over in heavy rain.

Arthur was a patient man; quietly reserved and deeply religious. He accepted that we had priorities for our income and that road-making was low on the list, but as the months went by he could stand it no longer, and offered to tackle the job himself with the aid of his two teenage sons, Granville and Ian, 'if I had no objection!' He agreed that I might give the boys some pocket money, but not a real wage. I had never before seen such enthusiasm applied to a job. With prodigious efforts, they cleared the stream inside the wood, transferring tons of beautiful soil from the wood and ditch to our garden, and rebuilt the drain under the lane, which they restored to a very respectable state. The postman was delighted, for he could now ride his bicycle all the way to our house. With grateful thanks to the Barretts, we began to use this 'top road' as our main drive.

Arthur's eldest son, Granville, became an expert collector of antiques, which he lovingly restored. In time, he renovated his beautiful Grade 1

The upper lane in winter presented a lovely scene.

Elizabethan house at Braisty Woods, then little more than a ruin, and developed a wonderful garden. Both he and Ian were sons to be proud of and gave us help and advice of every description, including winemaking equipment that set me off on an entirely new hobby when I retired. They, and their families, became lasting friends of ours. We felt particularly honoured because, traditionally, dales folk do not take to incomers quickly. My Yorkshire origins helped, but Margery always had a vague feeling that her southern accent got in the way.

Nevertheless, she made some firm friends, not only with Marion Barrett, but also with Mrs Hurley, wife of the village grocer, and the local butcher Tom Newbould. He had a small slaughterhouse behind his shop and provided us weekly with oxheart and fresh cows stomach, known locally as 'paunch'. The contents of the paunch were partly digested herbage, which suited our compost heap, but the paunch itself was dreadful stuff to handle—one couldn't rid one's hands of the smell, even wearing gloves, without the most vigorous scrubbing. However, the dogs loved it and it was very nutritious for them.

Our only serious clash with the locals came abruptly, and not long after our arrival. One Saturday morning, we found three men standing in the middle of our big field with guns at the ready and dogs at heel. We were aghast. How dared they trespass so presumptuously on our land? It was a severe shock to our treasured sense of privacy and peace. Margery was terribly distressed, and I had never been so angry in my life. I stalked up to the nearest man and demanded an explanation. He was embarrassed and apologetic, and took me to meet the captain of the shoot, Harold Atkinson, who owned the big mill at Glasshouses. Harold's first words, 'Now, don't get me wrong', became very familiar to us. This was his standard phrase for dealing with embarrassing situations. He protested that 'We've *always* shot over Woolwich, because it is at the centre of the shoot'. It hadn't occurred to him to ask our permission. He assured us that as landowners we had the right to join the shoot without paying any membership fee, and we would always be given at least one brace of pheasants after every shoot. Thus mollified, we not only allowed the shoot to carry on, but also over the years found ourselves thoroughly embroiled in rearing, feeding and watering hundreds of young birds. This involved building and repairing feeding pens, clambering over field stone walls with heavy buckets of grain and carrying them through the woods, often in wet and snowy weather. We certainly earned our share of the spoils.

I hated the sight of birds shot down, but came to accept this as a form of farming in which most of the birds we reared from chicks won their freedom, and had a good life. Without this sport, pheasants would probably have become extinct in the wild. Bob Myers was of like mind; he carried a gun but preferred the position of 'beating' next to me. His black Labrador, Pepper, was the spitting image of our dog, Salty. In dog training Salty had proved to be rock-solid in obeying commands, but when in the middle of a wood, with beaters shouting, birds rising, guns firing and birds falling, it proved too exciting for her to listen to me. Both Salty and Pepper had the bad habit of running in and Salty often brought me a bird, to the extreme annoyance of the

Chapter 9: 1962–1992 Woolwich Farm

Granville, on the tractor, was very patient with the rest of the shoot.

man who had shot it and whose dog should have had the privilege of retrieving it. In such cases, it was convenient for me to assume that the dog at fault was Pepper, and I would continue to call for Salty to come to me.

Bob adopted the same tactics with his dog, and so nobody ever knew who the culprit really was. Every year, Bob would bring me a sack of 'mucky carrots' freshly dug from his farm with instructions for storing them through the winter by covering them with sand. In beating through the woods, Bob and I were supposed to keep ten or fifteen yards apart, but one day when our paths converged under a tree, Bob put his finger to his lips and pointed upwards. There, in the crook of a branch only a few feet above our heads was a cock pheasant looking down at us, and staring with even greater concern at Salty and Pepper, who both had his scent and were casting around the base of the tree in an effort to find him (of course, the bird was not playing fair, because pheasants are never supposed to perch in trees). Bob and I chuckled and left him in peace. It pleased us both enormously.

I wasn't much of a marksman, and one day when we were beating up a steep hill towards Braisty Lane towards the line of standing guns, Salty flushed a pheasant that got wind of the guns and turned back to fly over our heads. This was precisely when I was supposed to shoot it, and I was just about to pull the trigger when there was a bang from the wood to my left and the bird fell at my feet. Salty retrieved it and when we were back at the farm watching the 'bag' being counted and distributed, Granville came up and congratulated me on my good shot. Well, I hadn't the slightest idea who had really shot it, but I did not argue with him. I muttered a 'thank you' and went home relieved that I did not have to do the annoying job of cleaning my gun!

At the end of each season, we beat the bounds of our territory to drive what few birds we could towards the feeding pens, in which we could look after them during the summer. On one such day, when Salty had not had much to do and must have been bored stiff, I suggested to Margery that I might take the only pheasant we had brought home that day and drop it so that Salty could make a retrieve. I put it under the magnificent old oak in our wood and Margery sent Salty to fetch it, but it was gone! My back had only been turned for a few minutes, but someone had been very smart—perhaps a fox—and that was the last of our supper. I once followed a fox all the way along the lane from Owen Well. It took no notice of me and slipped into our wood just behind our barn; a most beautiful creature.

Salty proved to be a resolute retriever, looking for a fallen bird long after all the other dogs had given up, and would astonish everyone by finally finding it. From her litter I selected a beautiful bitch, Bridie, loveable and intelligent. Full of hope, I took her to field training sessions together with Margery and her puppy William. Bridie was so successful that the trainer switched her to the advanced class, during which blank cartridges were fired as a cloth dummy was launched into the sky. The gunfire was totally unexpected and Bridie shrank in terror. For several weeks the trainer went out of his way to tempt her, by using dead birds and rabbits with far more interesting smells than dummies. Margery and I did everything we could think of to restore her confidence, but to no avail. She had become 'gun shy' and was ruined as a shooting dog. This was a great pity, for Bridie was exceptionally bright. For example, her mother Salty regularly pushed doors open, but could not be persuaded to shut them afterwards. Bridie could not only shut the sitting room door, but could *pull* it open!

We were dumbfounded when she first did this, something we never dreamed of teaching her to do—she had watched us and worked it out for herself. It required skill and effort, because the door was heavy, and the lever type handle required a firm downward pressure before it could be pulled open. To reach the handle she had to stand on her hind legs and put one paw against the door jamb, not on the door itself. She was rarely successful at her first attempt, but she persisted until she got it open, to her very evident satisfaction and to loud applause from everybody who was watching.

During a shoot, she had to stay quietly at home while the guns went off all around, but she enjoyed walks on the moors and in the woods. The first time she brought me a glove I had no idea that I had dropped it. She was so pleased at my reaction that it became routine for me to be terribly careless during our walks and regularly to lose hats and gloves, which she enthusiastically retrieved. The problem was how to 'lose' them without her seeing me drop them.

Bridie died early, but Salty outlasted everyone and needed a new companion. Our policy was to have at least two dogs, and we drove to Shropshire to buy Minnie, who came from a line of shooting dogs with impressive pedigrees. Training her was easy, and bearing in mind our experience with Bridie it was many months before I even assembled my gun. Minnie had never seen a gun before, but the moment the barrels clicked into place she sat up with ears pricked; alert in a way that simply astonished us. Her instincts were inherited from her marvellous forebears.

Margery and I were both now at an age when beating through woods over fallen trees and through thick undergrowth was too physically demanding, so we decided to reserve Minnie for the retrieval of fallen birds. On her first outing, we waited for the shoot to move on and then asked Harold if there were any wounded birds to be collected. 'Yes', he said, rather hesitantly, 'there's three, but they're rather a long way away. There's one over by Laurie Gill's bottom wood, another a'top end of your woods, and 'eaven knows where t'other one landed'. He pointed in three directions, and the instant Margery let go of the leash Minnie was off like a rocket. Some ten minutes later she was back with a bird; then she was off again, back with a second, and finally, after what seemed ages, she came in with a third!

It was such a thrill—we were as delighted as Minnie, and proudly caught up with the shoot and handed the birds over. Harold just could not believe it! He suspected that Minnie had gone to one of the pens and killed some live birds. He carefully inspected them, and only when he found the pellets that had killed them, did he apologise and congratulate us on her incredible performance. After that, Minnie never failed to find birds that the other dogs had given up for lost. She was certainly the finest retriever the shoot ever had.

It was years before, when we were living at Pasture Cottage that Margery discovered riding stables at Boston Spa, and coaxed me into taking some riding lessons. She rode a huge horse named Rupert and one day challenged me to a race. My mount, Big John, was even bigger than hers and when he fell behind Rupert, I decided to use my crop. Big mistake; I was just beginning to raise my arm when he swerved so violently that I only just managed to grab his mane and hang on by the skin of my teeth. This could only happen to a rank amateur, and Margery was determined that Sue should learn to ride properly. The riding instructor maintained that if you did not begin to ride *before* the age of five you would always ride like a sack of potatoes! Sue started riding lessons at Menston at the age of four, to avoid this fate.

At Woolwich we had two fields, with a beautiful fresh-water drinking hole in each, so it was inevitable that Sue was to have her own pony. There were a couple of small stables in the back yard, and all I had to do was to ring the fields with a line of wire above the stone walls, just to be sure, for old dry-stone walls have a tendency to collapse when pushed. It was just as well as I did. Sue's first pony, aptly named Speedy, tore round the field at frightening speed, looking for ways to escape. He was promptly replaced by Mandy who was downright lazy, but at least she was safe.

A problem arose; how could we exercise Mandy while Sue was away at boarding school? As usual, my resourceful wife had a solution. Having had the foresight to see that I could ride, she could buy me a horse and I could take Mandy on a leading rein when Sue was not at home. 'Mancia' was a somewhat irascible old skewbald that had been mistreated at the riding school. She accepted her new home gracefully, but made it clear that she did not intend to *work* in her retirement! While going on hacks at the riding school she had followed other ponies and jumped quite nicely over a gap in the hedge, but when I set up some small obstacles in our field she evidently thought that jumping over them would simply be a waste of energy. Even when I lifted my crop (I never actually hit her after my experience with Big John), she would refuse to jump and would carefully walk round the obstacle to prove her point.

On the frequent occasions when she didn't feel like going out, she would use all her wiles to resist being saddled and bridled, and then would do her very best to throw me. One day she bucked and cavorted for a quarter of a mile all the way down Mr Gill's field. When we finally returned home, Margery remarked how pleased she had been when she saw me lean over and pat her. I had to disillusion her; what actually happened was that Mancia had made one last desperate effort to buck me off, and she had damn near succeeded—I had to grab hold of her neck to avoid going right over her head. One day we started out by riding along the river bank and found it blocked by a pile of clippings from a hedge that was being trimmed. It was not much of an obstacle—maybe nine or ten inches high—but

Mancia liked a rest on our rides.

for once there was no other way round and I thought we might jump it. Mancia evidently thought otherwise.

I put her into a canter and we approached the 'jump' in fine style until, at the very last minute, she refused and threw me off sideways. So, in accordance with the instruction books, I remounted and tried again. This time she was tactful; the canter slowed to a walk, she stepped delicately over the twigs with an air of distain, and trotted on with the air of having done a good job. I was very relieved to think that no-one else had seen this pathetic episode, and we went over the bridge for a pleasant outing. Our return brought us back along the same riverside path. I was tired, Mancia sauntered at her own pace while I was dreaming of a nice cup of tea by the fire when I found myself suddenly flung high in the air and then dropped back into the saddle with a jolt that shook every bone in my body. Mancia stood stock still while I looked all around to see what on earth had happened. It couldn't have been a bite from a horsefly, or a wasp sting, for she would have bolted. It finally dawned on me that this was where the obstacle had been, for there in the middle of the path was a single twig with a few leaves on it, all that was left from the previous pile.

Of course, I had to congratulate Mancia on her very successful jump over the twig; it must have been on her conscience all afternoon, and we both went home content.

I think the happiest times Sue and I had on horseback were when she came home from school and mounted on Mandy, and I on Mancia, we trekked across the moors to Riva Hill, or to Rabbit Hill, or as far as Birstwith where the farrier had his smithy. These days blacksmiths arrive by van with a portable smithy in the back, but in those days you had to take the horse to the smithy. John Stockdale was a delightful man, with the typical sturdy build of a blacksmith, and it was a pleasure to watch him at work. He did not only shoe horses, but made many other things from wrought iron. The ponies felt better when freshly shod, so everybody enjoyed the outing and Mancia was always good on the ride home.

The fields provided ample grazing in the summer, but, as autumn approached, we realised that we must stock up with hay, for horse nuts from the chandler's were not enough on their own. To our dismay we found that, deep in the countryside as we were, hay was not at all easy to buy. Farmers tended to keep reserve stocks in case of bad weather, and as Mr Gill pointed out 'T'wer as good as money int' bank—and why didn't we mack us own?' I thought that it was probably more easily said than done, and I was right.

We did not make much progress in the art of haymaking until we met Mr Pearson, an elderly retired farmer who occasionally helped out at Braisty Woods Farm and who the Barretts highly recommended. He was a gentle man and a hard worker, in spite of his age, and he helped me every Saturday for years. He was always willing to give advice on haymaking, animal husbandry, timber, tractors, and farming in general. He enjoyed the splendid lunches given to him by Margery, but he did not reckon much to gardening—her favourite hobby—and would only condescend to help her in the garden as a very special favour, as it interrupted his 'more important work'. He had a relaxed outlook on life, and when we were worried about something he would reassure us with his favourite saying 'It'll tak ne fault'.

He advised me to buy an ancient Massey Ferguson tractor, which became my essential 'work-horse' for road mending, logging, building, and hay making with the help of a grass-cutter that Laurie Gill sold to me second hand. I suspect he was glad to get rid of the damn thing that was so awkward and heavy that, once mounted on the tractor, we never took it off again. It served well enough for many years, although it could not deal with anything near the edges and corners of the fields. So I had to buy and learn to use a scythe, to garner all those areas the cutter-bar could not reach. Before I could even start any actual haymaking, the field had to be combed to remove pebbles and stones that could break the cutter blades, all ragwort (lethal to animals when mixed with hay) had to be pulled up by the roots, and the whole field sprayed with weed killer. When I came in, quite exhausted, I had to persuade myself that it was good exercise.

Once the grass was cut (June hay is considered by far the best), our main problem was to keep it dry until it was baled and stacked indoors, because contractors with baling machines were busy with their major customers. After any rain, we had to keep on turning it to prevent it going mouldy. If hay bales are not bone dry when stacked, the moisture causes heat to build up and the hay can catch fire. Most barn fires are attributable to this. Since I was at work during the day, this arduous work had to be done during the evenings and at weekends and it always seemed to demand our attention during Wimbledon fortnight. It is fortunate, in a way, that in England the evenings are light until about 10.00 pm in June, so we could continue turning the hay while the light lasted.

Sue, then aged nine or ten, loved haymaking and so did the dogs. The Barrett brothers helped me buy a box for the back of the tractor to carry bales of hay (and Sue) down from the field. This was fun, but the 'Fergy' was a very lightweight tractor, and the box was designed for much bigger and heavier post-war machines. So, when full, it tipped the tractor backwards and lifted the front wheels off the ground. A huge lump of concrete between the front wheels was not enough to balance things, so I had to adopt the ludicrous posture of leaning over the bonnet to make the front wheels touch the ground. In this amateurish fashion we did manage to get a hay crop in, only to find that we had nothing like enough space to store it all under cover, and that became the stimulus for us to build a barn.

We had no money left to employ a building contractor, so I set about designing a barn that I could build in the evenings and at weekends. Local farmers were erecting steel framed sheds with corrugated asbestos sheeting for walls and roofs, but I had far too much reverence for Woolwich, with its beautiful old stonework, to commit such sacrilege. It had to be of stone to match the old buildings, with dressed quoins for all the openings and stone copings on the top of the walls. I was thinking of quite a sizable building and a pitched roof made of oak trusses would have been out of scale with the old house. So, I designed timber beams with a shallow pitch, supporting lightweight slabs, which would be sufficiently discreet to blend in with the older buildings without offence.

It was here that Mr Pearson came into his own; an expert dry stonewaller, he agreed that using mortar made building even easier. He found a small cart for sale in Pateley Bridge, and this was invaluable when I scavenged our locality for stone. A farm at Wilsil yielded some choice quoins from derelict outbuildings. The owner was glad to let me have them without charge—as long as I cleared the site completely. This was no problem, as I needed small stones for infilling. New York Mill, in Summerbridge, had a stack of huge stone slabs that they wished to get rid of, and the barn walls gradually grew to a height where we needed scaffolding on which to work. The cost of hiring scaffolding was prohibitive for a job that would last for months, but would be idle five days out of seven. So I got the local foundry to cobble together two steel frames that would straddle the wall and hold scaffold planks at a comfortable height for working. This contraption would never have worked in a traditional situation, but in our case, each layer of newly laid stones had five weekdays for the mortar to set, so it could support the scaffold as it rose one stage higher each week.

At that time, the hay-loft was still empty, and there was just enough space to make the roof beams, 24 feet long, from 4 x 2 inch battens, heavy plywood, and lots of galvanised nails and glue. It sounds amateurish, but the beams complied with the latest practice endorsed by the Institute of Structural Engineers, devised to offset the shortage of steel that was still only available with a license. The beams were light and incredibly strong, and Mr Pearson and I had no difficulty in positioning them, but I hired a couple of local labourers to lay the roof slabs. It was just as well that I took that precaution because it was tricky work, and one of them lost his balance and fell off the top of the wall—just managing to land on his feet, thank goodness.

THE 'WILSON' PORTABLE SCAFFOLD

The barn took more than two years to build. It was big enough to hold two cars, the tractor, tons of firewood, and all our hay, leaving plenty of covered space for chopping wood, mixing concrete, and so on. It meant that we could embark on modifications to the house and outbuildings that were to keep us busy for the next fifteen years, and well into my retirement. It was all hard work, but the woods, fields, and views across the valley were a constant joy as we built. Curlews would glide to the ground with their beautiful burbling warble, and peewits called from their nests in the adjoining field. The cuckoo would be calling, and thrushes and blackbirds would be singing in turn their mellifluous songs. When they stopped, the starling, whose nest was in the wall above the dairy, would chime in with its imitation of their songs, so there never seemed to be a moment when music was not in the air.

We had no accommodation for visitors, but the Leeds University Rover Scout crew occasionally came at weekends to give me a helping hand and incidentally to develop their skills in field craft and orienteering. On one occasion we agreed to meet at 'the rocking stone' in the middle of Blubberhouses Moor, which they proposed to approach from the south, and I from the north, to perform an investiture ceremony.

It was a cold, dreary day in winter, and Mancia was in a fractious mood that was exacerbated when she sank up to her hocks in the middle of the moor as we approached the rocking stone. We waited, and waited, until I became quite worried about the fate of my Rovers who were evidently lost on this desolate moor, or had perhaps suffered an accident of some kind (of course there were no mobile phones in those days—and no call boxes for miles on a moor). As it grew dark, I had to abandon the meet, to Mancia's very evident satisfaction, but I had waited too long, and with failing daylight and a mist beginning to fall, we were a danger to other traffic on those winding country roads until, bless her heart, Margery appeared in our car to escort us home, with headlights lighting up Mancia's white rump very effectively! Days later, we learned that the Rovers had had an equally anxious time waiting at *another* rocking stone on Blubberhouses Moor, miles away from mine (because of this, the whole crew and I came to thoroughly appreciate the value of map references).

In 1964, the Boy Scout movement announced the disbandment of all Rover Scout crews, and I decided to retire from Scouting after thirty-three years' service, including fourteen as a Rover Scout Leader. But the Rovers had other ideas and decided that the Scout and Guide Club needed revitalisation and would I please become their President? In vain I sought an excuse—surely they could find someone else? The university building programme was demanding a prodigious effort from me at that time. They were persistent and assured me that they would no longer depend on my help except for recruiting speakers. I should have known better; would I ever learn?

The Club found me useful in planning outings to the Lake District, and so on, and so I suggested something nearer home—how about a midnight hike across Blubberhouses Moor, ending up at Woolwich Farm? I lent them my prismatic compass so they could use 'dead reckoning' across the moor. I then pointed out that they would be arriving in the middle of the night, that isolated farms like ours were scattered all over Nidderdale, and that none of them had a name plate. I found it difficult to describe our house and then I had my brilliant idea; I would tie one end of a cord to my big toe and hang the other end out of our bedroom window, so that when they arrived and found the cord they would know they were at Woolwich, and a gentle tug would get me out of bed so that I could show them where they could sleep. Their trek over the moor was uneventful; the trouble only started at a farm on the edge of the moor. They were half way across a large field when they were chased by a herd of aggressive young bullocks. Then a watch dog raised such a shindy that the farmer came out with his twelve-bore shotgun and demanded to know 'what the bloody hell they were doing on his land at three o'clock in the morning?' In the meantime, I had been comfortably asleep in bed until, when I turned over, I felt a tug on my toe. Unfastening the cord and going downstairs I found no trace of visitors.

Once more in bed, the same thing happened again, and again. By the time I managed to adjust the cord so that it did not snag so easily, my visitors did arrive, but I was so profoundly asleep that they almost dislocated my toe before I accepted that this was the real thing and that they had finally arrived! However, the new barn provided enough hay for their comfort for the remainder of the night, and after a breakfast of sausages, eggs, and bacon they made their way back to Leeds. That was the last I saw of the University Scout and Guide Club and my compass, and it was the end of my scouting career.

Were there any detriments to living at Woolwich Farm? Well, power cuts were a nuisance, as we were at the very end of a very long service line, and had the lowest priority. Farms with cows to be milked understandably had high priority. One night, I awoke and saw with some fascination there were pretty red lights revolving on the bedroom ceiling. Very unusual at 4.00 am—or indeed, at any time! Looking out of our bedroom window there was a large fire engine and crew trying to douse a fire in Gill's barn. The barn had originally belonged to Woolwich farm and so was only twenty yards away. How the fire engine had managed to get there was, I suppose, some credit to my road-building abilities. They managed to put out the fire but the beautiful stone slate roof was gone, and from then on we gazed out at one made of corrugated asbestos cement. Sadly, it later transpired that the fire had been started by a member of the fire department; arsonists were an occasional hazard in many fire departments.

The old barn with its new roof, seen here with our granddaughters.

Being on a property with two natural streams could be problematic—one rainy morning we awoke to the loud sound of rushing water (rather than the usual trickling stream) and found that the garden stream had turned into a torrent forceful enough to move heavy wall stones out of its way. Then there was the winter weather and the challenge of snow.

One day it began to snow so heavily that the university staff were told to go home early.

All went well enough until I reached the top of Menwith Hill, to find a gigantic snowplough blocking the way over the 'tops'. 'Good', I said to myself, 'this will clear the worst bit'. But no. The driver got out of the cab and apologised, because he was stuck. The plough would only work when going forward, and this was an enormous American

Chapter 9: 1962–1992 Woolwich Farm

The torrent that awaited us one morning in the garden.

plough, used by the Americans at their US Navy Interception Station on Menwith Hill. He had begun clearing the nearby roads as a favour to the locals and now he was stuck in a lane so narrow he couldn't turn round. He was waiting for a gang to come and dig him out! So, I turned back and drove through the worsening storm back to Harrogate

We loved living at Woolwich Farm, here shown as it appeared in 1992.

and then up the dale to Summerbridge, past a snow-clearing gang who were amazed at my little mini coping with such conditions so successfully with its unique front wheel drive and with skid chains on. My tractor, with its huge rear wheels, could be used to clear the way down to the village with its hay box used as a snowplough—but unlike the American snowplough, my 'haybox plough' would only work when being driven backwards.

We loved living at Woolwich Farm, but sometimes it was forcibly brought to us that we did live in a very remote place. Our son-in-law mentioned this, as he drove his parents from Heathrow up to Leeds on the M1 (a six-lane motorway), turned onto the smaller dual carriageway from Leeds towards Nidderdale, from there to a one lane in each direction B-road, to a (to American eyes) road big enough for one car, to our lane, which dead-ended at our home.

Chapter Ten

1992–2018: A New Life In a New World

'There's a brand-new drug I can recommend, and it's not addictive', was the enthusiastic advice from the eminent neurologist when I complained that traditional painkillers were not having the slightest effect on my dreadful headaches. Little did he know.

Admittedly, Valium kept me at work for several years after this interview, but only at the cost of increasing the dose until my doctor warned me that it was getting dangerously high; he was proved right. During my routine work at the university, there were several episodes when I found myself flat on the floor. These events seemed inexplicable; I didn't feel unwell, and got up again without anyone knowing about it. I suppose I was just lucky that they never precipitated me down a flight of stairs, or occurred when I was driving. When I collapsed in full view of all the members of the University Planning Office, it was my dear colleague, Fred Wilkinson, who realised that something was seriously wrong. He called the medical officer, who asked Fred to drive me home.

The medical consultants insisted that I must not have any further contact with the university, and months were to pass before I was deemed well enough for Lord Boyle, the Vice Chancellor, to give me a farewell dinner in lieu of the usual retirement party.

What a way to end a career! To leave all my friends and colleagues so abruptly after thirty-four years without saying farewell to any of them! All my technical books, diaries, and papers had to be abandoned and left in my office cupboard, never to be seen again, for the medics warned 'You've got to forget all about the university and find other things to do'. The withdrawal symptoms were not too bad, and the new drugs transformed my usual nightmares into pleasant, happy dreams, of the sort I had not had for many years. A quiet life with Margery, the dogs, the goat, the hens, and the ducks was going to restore my health nicely.

Being retired at the age of sixty, and happily settling down to an idyllic retirement at Woolwich Farm, Margery persuaded me to enrol on a weekend painting course at Grantley Hall, a seventeenth-century mansion near Fountains Abbey. I do not know where she got the idea that I might paint; my efforts were pathetic, and I came home thoroughly depressed—I didn't have a clue. In spite of my protests, she insisted that I should make another stab at it. It transpired that Angus Rands, a local artist, whose work we both admired, held painting courses in the Yorkshire Dales, and she sent me off to one of his sessions at Reeth, in my beloved Swaledale, where I had so often camped with the scouts. After that course and another similar one at Kettlewell, I began to get some glimmer of how painting worked.

When Angus began to produce superb pictures in pastels, I decided to try my hand at everything. Oil paints took too long to dry, tested my patience, and were beyond my skill; the smell of turpentine was my excuse to abandon them. An extra mural course on acrylics at Leeds University irritated me; the pigments dried quickly and caught me by surprise, spoilt the picture, and ruined my brushes. Pastels were expensive, and I just could not cope with the mess. However, I could find no fault with watercolours, clean and quick, and in spite of their formidable reputation for being 'the most difficult medium to master', I decided to persevere with them.

Margery and I then went on a weekend course held by Ron Ranson in South Wales, which confirmed my decision to concentrate on watercolour. Although Angus was an accomplished artist, charming, and helpful, he had not been a good teacher. In stark contrast, Ron loved teaching; he was a flamboyant showman, and demonstrated what he called 'watercolour fast and loose' in his large studio in the Wye valley. He introduced us to the Japanese Hake, with which, together with a one-inch flat and a rigger, one could paint almost anything with

Chapter 10: 1992–2018 A New Life In a New World

incredible speed (these three brushes are now my standard tools, and Ron has become a prominent figure in the international art world).

At last I had the temerity to join the Harrogate and Nidderdale Art Club, which held out-door exhibitions in the Valley Gardens in Harrogate. I began to show pictures at village fairs, and I shall never forget the thrill when someone actually bought a painting at Masham, although I suspect they were motivated to support the local charity rather than by any appreciation of my art. Weekly club meetings were friendly and stimulating, with experienced members giving demonstrations.

It was exciting when the British Watercolour Society took over the Ilkley Art Show with a printed catalogue and prizes. Contributors came from all over England, and I hung my efforts with enthusiasm, only to discover that other artists were way better than me. Year after year my work was unsold and I came home in a state of despair.

Two annual holidays followed at the School of Painting at Innemore, on the Isle of Mull. These were marred only by Margery's appalling encounter with a bad oyster, and the dogs being sick after eating rotten carcasses from the moorlands over which they and Margery roamed every day. My confidence gradually increased as my technique improved, and I was able to produce much better paintings. I was pleased with my attempt at the packhorse bridge near Middleham in Yorkshire, and also with the old mill near Southwold in Suffolk.

When David Hinchcliffe opened his framing shop and art gallery in Pateley Bridge, he found that my pictures of local scenes were marketable, so he framed and sold them for me. As David was a talented artist who specialised in exquisite miniatures, this was most flattering and extremely convenient, and I even managed to win the Best of Show award at the Annual Exhibition in the Harrogate Public Library.

Thus encouraged, I entered the annual Laing National Competition held in London, at which my paintings were accepted, but not sold. In the third year, I thought that my entry, showing men fighting a moorland fire, was just too good to sell and I quadrupled my normal price to make sure I got it back, but it sold! I pocketed the money gratefully and made a replica, not half as good as the original was. That's the trouble with watercolour; the accidental blending of pigments and water just cannot be reproduced at will.

My confidence increased and I painted the packhorse bridge near Middleham.

Meanwhile, now as an Associate of the British Watercolour Society, I exhibited alongside the best at Ilkley. It had taken me ten years to crawl laboriously from the ground floor of this exhibition hall to the upper gallery. Nothing in my whole career had given me such satisfaction, and the crowning point came when I saw my name listed as a Fellow of the International Guild of Artists, which was limited to twelve selected watercolour painters. It was simply unbelievable. And to whom am I indebted to all this? Why, Margery, who inspired me in the first place, who would not let me give up and who allowed me to use her maiden name to sign my paintings 'Fawcett Wilson' as my tribute to her.

Sue rang from her home in Highland Park, near Chicago, which at our ages we never expected to see again. She was very excited and explained that we had a once in a lifetime opportunity to buy the house next to hers! 'What!—and leave Woolwich? She could not be serious!' But she argued, very tactfully, that we had no other relatives to depend upon, and as we became really old it would be difficult for her to keep flying over the Atlantic to deal with our problems. Moreover the house was a 'ranch type', with one main floor (ideal if we ever had to have wheelchairs) and only two bedrooms—it was perfect for retirement.

Chapter 10: 1992–2018 A New Life In a New World

There was only one major snag—we had to make a decision within the week. The owner had already received offers for the house, and Sue had been obliged to beg him to allow us time to make our own bid. Incidentally, she said, in the States such bids were binding and the deposit was non-returnable.

It was evident to me that if we moved it would be the end of my painting career, for the Yorkshire scenery was what inspired me the most, but that was nothing compared to the abandonment of our beloved farm, which we had worked so hard to create, and was now so gloriously comfortable. To leave the hens with their scrumptious eggs, the goat with her milk, the ducks, and all the wildlife we loved so much, would be hard. Harder still would be leaving Margery's garden that was now maturing to become a place of entrancing beauty in all seasons. In fact, Granville, himself an outstanding gardener, confessed that he used to sneak along and sit in our garden when we were away. He reckoned that he had a proprietary interest, having donated loads of lovely farmyard manure and with Ian, diverted a field drain and built a dry stone wall at the top of the garden, where he liked to sit and reflect.

We did not sleep at all that night, and not much on subsequent nights, for we had to admit that Sue had a point—if not several. We both suffered from arthritis; my hernias made lifting slow and digging slow and painful, and wielding a felling axe and operating a chainsaw was no longer possible. Even kneeling down to milk Nanny was difficult, and getting up even more so. Sue clinched the deal by pointing out that her three daughters would not get to know their grandparents very well if they only saw us once a year, whereas if we lived next door we could watch them grow up. So, we put all our trust in our darling daughter and authorised her to buy a house that we had never seen, and use all her persuasive powers to get the owner to accept our bid. Which she did.

At last, Sue was able to send us some photos of the property we had so trustingly acquired. It looked modest enough: two bedrooms, one combined living room and dining room, a tiny kitchen, a single garage (attached to the house—that was a novel idea), and some basement storage. I wondered if we would find ourselves short of space. The sheer volume of our accumulated paraphernalia in the farmhouse and outbuildings was horrendous, but we set about selling it. My model aircraft collection I offered at car boot sales, our books went to second hand bookshops in Harrogate and York, of which there were some excellent ones. My easels, drawing boards, drafting table and HiFi equipment all had to go, with all my workshop tools and our furniture. As everyone who has had to sell up knows, you try to get it right but inevitably you let go of a few items that you find you need later on, and you keep a few items that turn out to be totally unnecessary. We were spurred on by the weight and volume restrictions for moving overseas.

Our anxieties increased as we learned about the high cost of medical insurance and medication in America, for we had no idea what the sale of Woolwich Farm was likely to realise and the property market in the UK was in a slump. We had to pay cash for this new home, and I began to panic as I wondered if anyone would pay good money for an old house that could only be reached along half a mile of narrow, twisting, gravel country lane, impassable in heavy snow. To our amazement and delight we received three offers, way beyond our expectations. We accepted one, and the new owners told us that they had often walked past Woolwich and had determined that this was the property they *must* have, if it ever came on the market.

So, after many traumatic and busy, busy weeks, we finally arrived in Chicago together with our black lab Minnie, quite unprepared for the cultural shocks that awaited us. For a start, everything was so BIG! We were collected from the airport in a huge car, and on the way home we passed through a 'micro burst'—this was July—a violent thunderstorm that hurled debris and branches across the motorway. It was unbearably hot and humid, and we were glad to be, for the first time ever, in an air-conditioned car. We were delighted to be told that our new home had air-conditioning too—a luxury we had never dreamed of having.

When we took possession of our new home, we were in for a shock; it was far bigger than any two-bedroomed house we had ever come across. Who could have imagined a sitting room 36 feet long? The little kitchen was perfect though—a beautifully equipped galley with just enough room for one cook. In addition to the luxury of air conditioning, to my delight, I found that we had under floor heating, rare in America where warm forced air is the norm. Evidently, the house had

Chapter 10: 1992–2018 A New Life In a New World

been built to this unusually high standard by a builder to be his own home and the whole design was so quietly sophisticated and efficient that I should have been very proud if I had designed it myself.

Next, we had to buy some furniture and equipment, so we opened bank accounts only to find that shops refused to accept our cheques. They needed proof of identity such as a driving licence (not a British one, thank you) or a green card (requiring six months residence in the States). It would be at least a few months before I could get either of these, but thank goodness everybody said they would accept a credit card.

No problem. I had used an Access card in England for the past twenty years. In fact, back in the early 1970s, I had got one of the earliest credit cards seen at the University for an overseas trip, and my colleagues had made fun of it as a gimmick whose time would soon pass. Our credit was well-nigh perfect, and we owned our own house outright. However, every application was turned down, because I 'had not been in the country long enough'.

It was unbelievable—with no cheques or credit cards, every item had to be acquired though the agency of our long-suffering daughter, who was already busy with the complexities of getting our green cards, health insurance, and coaching me how to drive a car without a gear shift and designed for use on the wrong side of the road. We could not have managed without her constant help and advice, but it was challenging.

So was the weekly grocery shopping. The bewildering choice of bread was all too sweet, with indigestibly thick slices. No matter—we had always baked our own bread. Unfortunately, our usual ingredients of strong flour and plain flour were unknown here, and their all-purpose flour refused to work with our recipes. We could not make decent marmalade either, from a lack of the sharp Seville oranges. Our standard foods such as tripe, ox heart, kidneys, or calf's liver became history, as none could be obtained at the local supermarkets. Gone were the days when Margery would complain 'I'm getting tired of cooking pheasant and jugged hare!'

Eventually one of the butchers succeeded in finding some calf's liver, and delighted us by telling us that his wife 'had never tasted anything so good!' I also brightened up when I discovered the enormous local liquor store, with its unbelievable range of wines and spirits from all over the world. Back in Britain, taxes on alcoholic drinks had increased gradually to the point that we simply could not afford them and relied on our own homemade wines, with the luxury of a bottle of single malt at Christmas. Now, after some experimentation with exotic spirits, we could afford Scottish whisky and French brandy, which our doctor said was good for us.

Circumstances improved dramatically when I passed my driving test and got a licence, which meant I could use my cheques and take advantage of all the special offers, coupons, and returns policies with no questions asked. This was in stark contrast with the practice in Britain. Margery had once given me a Rolex watch, which had the disconcerting habit of stopping and then restarting by itself, so I kept finding myself late for appointments and once nearly missed a London train. After several attempts at repairing it, the very reputable jewellers in Harrogate refused to refund or replace it, and, later, Rolex admitted that its design had been faulty. I had thrown mine away. There were no gift cards or return policies so one had to be very careful when buying gifts—in fact homemade items were always the safest option.

In America, things were quite different and I began to like living here. I now understood why Americans enjoyed shopping, and why they continued to build vast shopping malls in their beautiful countryside. One thing did puzzle me—how on earth could they pay for the heating and cooling of these vast premises? The answer seemed to lie in the reduction of labour by the use of electronic accounting, to the point where some of these shops were so sparsely staffed that it was difficult to find an assistant to make a sale. Even in small shops things could be difficult, as when I asked in Radio Shack for batteries for my flash light: 'Sorry sir, but I can't let you have any—the computer is down'. 'But', I said, 'I have the exact money, I don't need any change from the till; I don't want a receipt, and I need the batteries urgently', but to no avail. The manager was summoned only to apologise profusely and to confirm that, without the computer, there was no way to make a sale.

Our local supermarket was also amusing. When it advertised a 10 percent discount on any purchase over $100, I loaded up my shopping trolley to the gunwales and at the checkout the result was totalled at $132.50. I was slightly embarrassed when I found it necessary to remind the cashier about the 10 percent discount, and even more so when he promptly knocked fifty cents off the bill and asked me for $132 instead. When I pointed out that 10 percent was a lot more than fifty cents he looked puzzled and rang for the manager, as a queue of people began to form behind me. The manager immediately conceded that fifty cents was not right, but said 'I'll have to go to my office to get my calculator'. The queue began to grow, and so did my embarrassment as we all waited for him to return. After two attempts with his calculator he agreed with my figure and told the cashier to reduce the charge by thirteen dollars. The cashier still looked uncertain, and it was necessary for the calculator to be once more employed to deduct the thirteen dollars from $132. To my intense relief the people in the line behind me appeared to regard the incident as entertaining rather than an annoyance.

Developments in the 1990s compounded our geriatric confusion, with the introduction of the internet, digital cameras, cell phones, and digital video discs in a bewildering and hectic cataclysm of new inventions. My old friends back in England agreed that the best thing was to let the youngsters get on with these new-fangled devices without us. But then Margery rocked my boat by asking for a simple camera for her birthday. To my dismay the only ones available were digital of course, with film being a thing of the past. Next, I was faced with my defunct video recorder, and had to cope with the new system of DVDs. So, I could not escape the blasted digitisation of the world, and have since struggled to operate in it with a skill comparable to that of the cashier at the supermarket. However, after twenty years of effort, it must be said that I am writing this on a computer, and that I do have modest email and Internet skills. There was no real escape possible.

In the meantime, I attended to the basement. This was far more extensive than we had thought at first, and after months of exciting work, dismantling heavy shelving and benches and breaking through an internal wall, I succeeded in creating a studio any artist might dream about, together with a framing workshop and a large picture gallery. Although all the art materials I had left behind in England would have to be replaced, I thought perhaps I should start painting again. Enquiries revealed that nothing like an English Art Club existed in the Highland Park area, but there were painting classes with fees varying from $100–$200 for a ten-week session! This set me back, for my old club in Yorkshire had met regularly throughout the year, provided demonstrations, and held an annual exhibition, all for a single subscription that cost the equivalent of about ten dollars a year. After a careful review of our pensions, Margery and I enrolled in a course on figure drawing at the local Arts Centre. The instructor arrived an hour late, explained that the fees were too low to afford to hire a model, handed round some photographs for us to copy, and went off to get a cup of coffee. We demanded, and got, our money back.

Then we heard about the North Shore Art League in Winnetka, where a distinguished watercolourist, John Dioszegi, used a technique that was completely new to me. He eliminated the awkward problem of 'cockling' by thoroughly soaking both sides of the paper before he started to paint. This appealed to me because it saved a lot of money. Hitherto I had used 300 lb weight, which was too stiff to cockle, and was expensive. John used best quality 140 lb paper, at less than half the price; a man after my own heart, and I promptly enrolled in his class. He specialised in abstracts, created by dropping generous globs of paint on to wet paper and then manoeuvring the board so that they blended with the pigments already there—a technique known as 'wet into wet'. No other medium could achieve such subtle colours and shapes, and it was fascinating to watch him.

John's pictures were admired and sold well, for they were very decorative. But when I emulated his work the results never really satisfied me, for they seemed to cry out for some further resolution. So, I adopted his basic technique, and then slowly groped my way towards some unknown concept, persuading the paints themselves to develop into a picture. Manoeuvring the flowing paint to form recognisable features such as lakes, rivers or foliage was far from easy. I once found myself faced with an area of blue and turquoise, flanked by yellows of various hues that hung in the studio week after week as I tried to decide how it might be transformed into a river, a lake or the sea. In every case the perspective was wrong. I finally took a 'bird's eye view' and converted the yellows into cliffs enclosing an estuary with a jetty and fishing boats. It was fun inventing this scene and 'Smuggler's Cove' is now a favourite painting.

With flowers, which had defeated me hitherto, the 'wet into wet' method came up trumps, but in the Chicago Botanical Gardens, I found the heat and glare more than I could cope with, and it was a relief when John held a course in Wisconsin. The weather had cooled down, and the humidity was, for me, back to normality for it had rained steadily so I was back in my natural element.

After that, events happened very quickly; John was unable to attend the show at the Wilmette Arts Guild and he asked me to take his place. I was warmly welcomed, sold some paintings, and happily joined the Guild. John then decided to take a sabbatical year and asked me to look after his classes at West Park and the North Shore League. I was dumbfounded, but he assured me I could do it, and I found him impossible to refuse, although I was sure I would make a fool of myself. It helped that I had been a student in that class, where most were mature students and all of the men were retired. None had much experience, so I felt a great deal of empathy with them, having been faced with learning problems myself so recently, and found no real difficulties in helping them. Demonstrations were a different matter; they terrified me. I made no attempt to emulate John's teaching methods,

Afghan Poppies.

but I was able to show them his ingenious method of dealing with the tricky business of tone values.

Paintings can prove unsatisfactory in poor light if the tone values are wrong. In a balanced picture, the tones preserve the integrity of the painting even if one cannot distinguish colours. With watercolours, imbalances in tone do not register until the paint is thoroughly dry, after which it can be difficult to rectify any faults. Even experienced artists can be caught in this trap, so John's solution was extremely helpful. He had a card with small holes punched in it through which one could view the tone of any questionable area without the distraction of surrounding colours. The tone value could then be assessed in relation to the white card, or compared to another hole, and this made it much easier to see if any amendments were needed to achieve a tonal balance.

I knew that demonstrations could be boring, so I kept them short, only using them to illustrate solutions to problems, but on one occasion I attempted a quick study. I was shown a photo of a snowy street in Montreal, and I was tempted to do a quick sketch showing what could be achieved by using only two pigments; French Ultramarine blue and Burnt Sienna brown. The resulting painting was remarkably successful and I think we were all impressed with what could be achieved in only ten minutes.

One difficulty came from language; understandable with students from Korea and Japan,

Demonstration using only two pigments.

Chapter 10: 1992–2018 A New Life In a New World

but more unexpected when dealing with native English speakers. When I found a student using a hard rubber I called the class together and explained how this would destroy the surface of the paper and ruin the picture before it was started. Of three types of rubber, hard, soft, and putty, I explained that only the putty rubber could be used with safety. There was a horrible silence, until one lady took pity and whispered to me, 'In polite society we use the term eraser'. I was completely mystified. I had to wait until Sue explained things to me, but after seventy years continual innocent use of the word rubber I found it very difficult to prevent it slipping out occasionally. In time, the class got used to my quaint accent and terminology, and ignored my verbal gaffes. I also got used to their terminology. Phrases I became familiar with were, for instance 'Jeet?', which means 'Have you eaten yet?' and another was 'Squeet?', which means 'Let's go and eat'.

Students were gratifyingly pleased by my efforts to help them, which was complicated by their inability to describe their difficulties in a coherent sentence. To achieve any precision in language, you need the right vocabulary and it took patience and perseverance on my part to understand their problems. I found that demonstrations, replacing words with visual representations, were by far the best way to help them. I usually came home exhausted but elated, and always wiser. At the end of each ten-week term we held the traditional 'crit' session where everyone, including me, brought work along to show and be criticised freely. To most people it was fun, and valuable, but others found it an ordeal and skipped it. I do remember how impressed I was at the quality of most of the work, and how delighted we all were to hear when some of their paintings were exhibited and sold.

It was my dentist, Gary Alberts, who first showed a serious interest in my work and hung it throughout the rooms of his dental office, giving me wonderful publicity. Shops in Wilmette who displayed the work of local artists supplemented the annual shows of the North Shore Art League and the Wilmette Arts Guild, and I was kept busy. John resumed his evening classes at West Ridge, but the North Shore League asked me to continue with my classes. Then several students asked me if I could hold a weekend workshop, and it was then that my studio accommodation came into its own, with room for fifteen artists to work indoors.

Teaching painting at the North Shore League was very satisfying.

This workshop was such a success that I was faced with a demand for regular tutorials, which I held three days a week. Then John gave me more to do—every year he had produced a picture for the cover of the local Park District catalogue, with his original work being sold in aid of charity, but John felt he was running out of ideas, and recommended me for the job. I was horrified; my experience of commissions had taught me to avoid them like the plague. When he said that this was an honour, I became even more nervous. The traditional subject was local landscape and Rosewood Park, which I knew best as it was where we exercised Minnie, was nothing but a bland mixture of green grass and green leaves bordering Lake Michigan. Then I remembered our winter walks along the edge of the lake with its dramatic shelf of ice and the incredible blues and turquoise colours of the water. My problem was solved, and the 1997 catalogue included a flattering biographical note and photo of 'the artist'.

The traditional league exhibition at Old Orchard was surrounded by noise, dust, and dirt from the construction of the new Nordstrom

Chapter 10: 1992–2018 A New Life In a New World

Rosewood Park, 1997.

store, and so the event moved to other sites. Carrying, repositioning, and arranging the art was heavy work, and I was grateful for help from Margery and Sue. The summer heat was always a problem, and finding shade in open parks was difficult, so when the venue changed to the air-conditioned Northbrook Court Mall, everybody was pleased.

Early the next summer, as I was taking some artwork upstairs, I fell backwards and hit my back against the wall. The following morning I did not feel well. Sue took me to the ER, where the cardiologist diagnosed a catastrophic heart attack. My fall had loosened a piece of plaque in an artery, which was now blocking the blood flow. The normal procedure was to insert a stent into the artery to improve the flow, and I was lying on a gurney being prepped for this when I enquired as to the cost. I had no medical insurance, and had not been in the country long enough to qualify for Medicare. The cost was estimated at about $40,000; I quickly asked if there were any alternatives. The surgeon decided to send a camera inside me to have a look and to his horror found that my heart had ruptured and was being held together by the pericardium. The procedure to insert the stent would have killed me!

This was a very rare condition; a study in Spain had found that out of nineteen cases, split roughly between operating and 'waiting and seeing', the outcome was about even; only half survived. After discussion with his colleagues, while I lay in the operating theatre, he apologised for keeping me waiting, explained the issues to me, and ended by saying 'Of course, it's entirely up to you to make the final decision—I'll do whatever you think is best'. I have never made a decision more quickly, and with careful nursing, my heart did eventually heal itself.

However, teaching was out of the question, and I had to explain to thirty-five students and the North Shore League why I had to retire. This was easier said than done—heart operations were considered routine, and most patients were back to work in days or weeks at most, so people could not understand why this was not the case with me. I received some very heart-warming letters that I treasure because they remind me of the brief and very happy time I enjoyed teaching them.

The Lookout.

Review many of GW's paintings in colour at GeoffreyFawcettWilson.com.

Epilogue

In retrospect, I was a serious child, and compressing both my school and university careers left not much spare time for fun. It seems that teaching was the key, and Prof Evans's invitation to join him in the creation of the Department of Civil Engineering lifted my spirits and gave me an indescribable thrill. However, the pressure of work was relentless, and my dream of becoming the first Professor of Architectural Engineering had to be abandoned when I yielded to the blandishments of the Vice Chancellor to join his staff.

My new job as University Planning Officer was onerous; whenever I prevented the university from making mistakes, others took the credit, whereas if things went awry, I was blamed. The struggles to extract money from the UGC and to avoid overspending were a constant worry. Things brightened up with the advent of the brilliant Chamberlin Plan but, even then, my troubles were not over, for there was opposition to Chamberlin being allowed to actually design any buildings. University staff criticised Chamberlin's 'brutalist concrete structures' and I had to fight battles to secure further commissions for him. In her book[1], Elain Harwood says of Leeds, 'As a landmark of university planning, it deserves to be better known'. It was a privilege to work with people like Joe Chamberlin and Frank Woods, and I became very fond of them. My professional relationships with other gifted archi-

---

[1] *Chamberlin, Powell & Bon* by Elain Harwood, RIBA publishing ,2011

tects such as Ed Hill of Building Design Partnership and Dennis Jones were harmonious. I had loyal support from the members of my Planning Office, and constant encouragement from Sir Charles Morris and later from Lord Boyle. I am grateful to them all, and realise that I have been a very lucky man to have worked alongside them.

I suppose my collapse was inevitable, but my subsequent life, making model vintage aircraft, painting in retirement at Woolwich and then teaching watercolour in the USA has been a marvellously happy and rewarding experience.

Both Margery and I have so much for which to be grateful, with several colleagues and friends still surviving, who have given us wonderful support, together with such a loving family. I hope this epilogue will be accepted as an expression of our gratitude to them all.

Index

Note: *Italic* page numbers indicate a figure.

A

Airie, Sir Edwin, 92
architects, university projects
    Airie, Sir Edwin, 92
    Johnson, Allan, 80, 104
    Jones, Denis Mason, 100, 103
    Knighton, A L, 102–103
    Lodge, T A, 79–80, 91, 93–94, 103–104
art exhibitions, 225–226, *226*, 235
Aston, Guyron, 188–191
Atkinson, Harold, 208, 211
Austin Mini, 147–148
author's works, *233, 234, 236, 237, 238*

B

Baden-Powell, Lord Robert, 20, 32, 44
barrage balloons, 62, *63*
Barrett, Arthur, 205, 216
Barrett, Granville, 206, 210
Barrett family, 205–207, 209, 215, 216
BDP consultants, 174–175, 179, 180
bicycle, 60, 66, 70
Black Badger, 38
Bodington hall development, 165–168
Bodington, Sir Nathan, 167. *See also* Bodington hall development
Bon, Christopher, 152, 172, 174
Bottesford, England, 1
Boy Scouts
    Everett, Sir Percy, 46
    Baden-Powell, Lord Robert, 20, 32, 44
    early years, 20, *21*
    high school years, Belle Vue
        Black Badger, 38
        camping mischief, 36–39
        cooking duty, 35–36, *36*
        Duke of Kent's Messenger fundraising appointment, 43–46, *45*
        Emmott, Ken, 38, 41–42, 51
        hazing, 33
        International World Jamboree in Holland, 42–43
        Lake District bicycling tour, 40–41
        Law, Jack, 40–41
        London Tube confusion, 44–45
        Red Dove, 35, 38, 46, 47, 48, 52
        Singing Hawk, 35–36, *36,* 37, 38, 45
        Switzerland trip, 46–50, *49*
        tent feed, 38–39
        uniform particulars, 32
        woodcraft, 34
        Woodcraft Circle, 35, 37–38
    University years, pre-war
        bridge building, 52–53
        fog patrols, 53–54

Boyle, Lord, 223
Bradford, England. *See also* early years
  early years, 1–2, 4–6
  tram travel, 29–30
Brimham Rocks picnic, 20–22
British Watercolour Society, 225, 226
Brittain, Dr. Robert, 123, 124–125
Brotherton, Lord, 87. *See also* Library
Brown, Bursar, 100–101
Brown, James, 84–85
Brownlie, Miss, 97
Burton family, 145–146
Butler, Reg, 138–139

C

camping mischief, 36–39
Chamberlin, Peter, 152–174, *170, 171, 172*. *See also* Chamberlin Plan
Chamberlin Plan
  accolades, 173–174
  BDP consultants, 174–175, 179, 180
  Bon, Christopher, 152, 172, 174
  Central Electricity Generating Station, 179–180
  Chamberlin Development Plan Book, 153–156
  Chamberlin's partners, 152
  Chamberlin's personality, 152–153, 173
  cost control measures, 173
  flexibility in building uses, 154
  Kaberry, Sir Donald, 174–176
  lecture theatres, 155–156, *171*
  library, 170–172
  Mathematics and Sciences building, 170, *172*
  medical school, 174–178
  Poulson, John, 157–158
  Powell, Geoffrey, 152, 172, 174
  ring road proposal, 154, 156, 158–159
  selection for T A Lodge replacement, 151–152
  student housing, 159–168
  Thirlwall, Geoffrey, 158–159
  topping out ceremony, 170
  touring other universities, 159–165, 177–178
  Wood, Derek, 177–178
Chapman, 77–78
charabanc, 2–3, *3*
Chicago Park District commission, 236–237, *237*
Chicago relocation. *See also* painting
  heart attack, 237–238
  painting once more, 231–237, *233, 234, 236, 237*
  selling Woolwich farm, 227–228
  shopping challenges, 229–231
  Susan's appeal, 226–227
  technology challenges, 231
chores, childhood, 7–9, *8*, 10
cinema, early years, 14
Civil Engineering department development
  Chapman, 77–78
  designing new Engineering building, 78–80
  Evans, Dr R H, 73–75, 78–79, 81
  Happold, Ted, 82
  lecturer duties, 74–75, 77, 80–82
  Lodge, T A, 79–80, 91, 93–94
  Maillart, Robert, 81
  Morris, Vice Chancellor Charles, 79, 79n2, 80, 82, 86, 91, 100, 102–104
  rationing challenges, 76–77, 91
  Rider, Sidney, 73–74
  Sutherland, Dr, 78
Clegg, Russell, 35, 38, 46, 47, 48, 52
conferences and consultancies
  asbestos presentation, 181–182
  Chamberlin Plan presentations, 182
  erratically evolved departments, 184–185
  Expo '67, 182
  KDF9, 184
  MacKinnon, Murdo, 182–183, 184
  MIT, 183–184
  Planning Officer position, university Arts Block, 137–138
  Botany department, 134–135

Index

Chamberlin Plan, 151–180, *170, 171, 172*
Friend, Louise, 101–102
Great Hall remodel, 100
Gregory fellows, 138–139
House and Estates Committee, 101–102
Library, 87–90, 170–172
Man-Made Fibres Building, 135–137, *136, 137*
new Refectory, 94–98
Parkinson Building, 91–94, *94*
support staff, 101–104
University of East Africa central catering unit
 arrival in Nairobi, 187–188
 Aston, Guyron, 188–191
 Glew, George, 186–187, 191–192
 Jones, Royston, 188–191
 Principal report, 191
 related problems, 192
 Tsavo Game reserve, 190–191, *191*
University of Leipzig tour
 arrival in Berlin, 192–193
 Crawford, Ron, 192–198
 Hans the tour guide, 193–198
 reciprocal England visit, 195–198
 restoration sites, 194–195
 Trabant road trip, 193–194, *194*
 Ullmann, Herr Helmut, 194–198
courtship of Eleanor Margery Fawcett
 engagement and wedding, 126–127
 first kiss, 126
 Woodsley Hall, 121–123, *122*
Crawford, Ron, 192–198
Crockatt, Helen, 143, 145

D

Dainton, Fred, 146
Denny, James, 92–93
designing new Engineering building, 78–80
Devonshire, Duke of, 98–99
Devonshire Hall residence, 83–86
 discordant dueling pianos, 84–85
 missed dinner engagement, 83–84
Dioszegi, John, 232–234, 235–236

discordant dueling pianos, 84–85
doctoral studies, 57
doodlebugs. *See* V1 flying bombs
Duke of Kent's Messenger fundraising appointment, 43–46, *45*

E

early years
 acting, 24–25
 boiler treatment business, 4–5
 Boy Scouts, 20, *21*
 Bradford, England, 1–2, 4–6
 Brimham Rocks picnic, 20–22
 chores, 7–9, *8*, 10
 cinema, 14
 Flies Right Off the Ground aeroplane, 27
 Grandfather, 16–17, 20–22
 Leicester Christmas visits, 26–27
 Lister's Mill, 6
 Malcolm, 6, 10, 21, 27–28
 milk delivery, 15–16
 Paddy, 17–19
 Peel Park, 6, 27–28
 piano, 19–20
 politics, 25–26
 Polly parrot, 16–17
 Priestley, J B, 25
 radio, 19
 Roslyn Terrace, 3, 19
 shopping, 9–11
 Smith, Arnold, 15–19, *18*
 toys, 27–28
 Undercliffe, England, 5–6, 9, 14, 28
 Wilkinson, Cecil, 22–26, *23*
 winter sports, 28
 Yeadon aerodrome, 13, 21, 27
Edinburgh, Duke of, 135–136, *136*
Ellerton Abbey ghost, 40
Ellis, Stanley, 124
Emmott, Ken, 38, 41–42, 51
entrance exam challenges, 29
epilogue, 239–240
Evans, Dr R H, 50, 71
 Civil Engineering department development, 73–75, 78–79, 81
 driving adventures, 74

— 243 —

Evans, Rhydwyn Harding. *See* Evans, Dr R H
Everett, Sir Percy, 46
Expo '67, 182

F

Father. *See* Wilson, Arthur J.
Fawcett, Eleanor Margery, 121–122
   engagement and wedding, 126–127
   first kiss, 126
   pregnancy, 127–131, 133
   thyroid surgery, 142–143
   Woodsley Hall, 121–123, *122*
Fighting Vehicles 9 department, 59, 61, 69
Flies Right Off the Ground aeroplane, 27
fog patrols, 53–54
forced retirement, 223–224
Foxton, Barry, 202–204
Friend, Louise, 101–102
friends, early years
   Pickles, Jack, 10
   Shaw, Malcolm, 6, 10, 21, 27–28
FROG. *See* Flies Right Off the Ground aeroplane
FV9. *See* Fighting Vehicles 9 department

G

Garton, Sam, 51, 53
Gerty the smuggler hostess, 129–130
Gill, Laurie, 213, 215, 220
Glew, George, 186–187, 191–192
Grandfather, 16–17, 20–22, 31, 33
Gregory, Eric, 138. *See also* Gregory fellows
Gregory fellows, 138–139

H

Hans the tour guide, 193–198
Happold, Ted, 82
Harewood, 22, 100, 143, 145, 168
Harwood, Elain, 174, 239
hazing, school scout troop, 33
heart attack, 237–238

'Henry Price' flats, 165
high school years, Belle Vue
   Ellerton Abbey ghost, 40
   entrance exam challenges, 29
   first bicycle, 30–31
   Malcolm, 30–31
   qualifying for university, 50
   school scout troop
      Black Badger, 38
      camping mischief, 36–39
      cooking duty, 35–36, *36*
      Duke of Kent's Messenger fundraising appointment, 43–46, *45*
      Emmott, Ken, 38, 41–42, 51
      hazing, 33
      International World Jamboree in Holland, 42–43
      Lake District bicycling tour, 40–41
      Law, Jack, 40–41
      London Tube confusion, 44–45
      Red Dove, 35, 38, 46, 47, 48, 52
      Singing Hawk, 35–36, *36*, 37, 38, 45
      Switzerland trip, 46–50, *49*
      tent feed, 38–39
      uniform particulars, 32
      woodcraft, 34
      Woodcraft Circle, 35, 37–38
   stealing a Bunsen burner, 39
   tram travel, 29–30
High Wycombe holiday, 70–71
Hilliard, Hall Porter, 98–99
Holford, Sir William, 140–141

I

Ilett, J J, 86–87, 101
International World Jamboree in Holland, 42–43

J

Johnson, Allan, 80, 104, 137, 179
Jones, Denis Mason
   Bodington hall development, 165–168
   car fire story, 168–170

Index

Great Hall remodel, 100, 103
Jones, Royston, 188–191
Jowett motor cars, 13

K

Kaberry, Sir Donald, 174–176
KDF9, 184
Kent, Duchess of, 100, *101*, 126, 160
Kent, Duke of, 43, 44, 46
King's Own Yorkshire Light Infantry, 1
Knighton, A L, 102–103
Knowles, Ernest, 35–36, 36, 37, 38, 45
KOYLI. See King's Own Yorkshire Light Infantry

L

'La Contessa,' 143
Lake District bicycling tour, 40–41
Lasdun, Denys, 151,174
Law, Jack, 40–41
learning in England, 224–226, *226*
lecture theatres, 155–156, *171*
lecturer duties, 74–75, 77, 80–82
Leicester Christmas visits, 26–27
Leipzig visitors, 195–198
Library, 87–90, 170–172
Lister, Miss, 133–134
Lister's Mill, 6
Lodge, T A, 79–80, 91, 93–94, 103–104, 135–137
London living arrangements, 59–60, 61
London Tube confusion, 44–45

M

Mac. See Shaw, Malcolm
MacKinnon, Murdo, 182–183, 184
Maillart, Robert, 81
Majorca holiday, 128–131
Malcolm
  early years, 6, 10, 21, 27–28
  high school years, 30–31
Man-Made Fibres Building, 135–137, *136, 137*
Manton, Irene, 134–135
Margery. See Fawcett, Eleanor Margery
Martin, Sir George, 156

Marmolada, 113–119, *115, 116*
Middleham castle, 196-197
milk delivery, early years, 15–16
Mont Buet, 105–113, *108, 109*
Morgan family, 144–145
Morley, Emma Wilson, 26–27
Morris, Vice Chancellor Charles, 79, 79n2, 80, 82, 86, 102–104
Mother. See Smith, Edith
mountaineering
  Marmolada, 113–119, *115, 116*
  Mont Buet, 105–113, *108, 109*
moving to Chicago, 226–231
'Mystery Tours,' 2–3, *3*

N

North Shore Art League, 232, 233, 235

O

Ode of Remembrance, 69
Ogilvie, Lady, 99

P

Paddy, 17–19
painting
  art exhibitions, 225–226, *226*, 235
  author's works, *233, 234, 236, 237, 238*
  Chicago Park District commission, 236–237, *237*
  Dioszegi, John, 232–234, 235–236
  heart attack, 237–238
  learning in England, 224–226, *226*
  moving to Chicago, 226–231
  North Shore Art League, 232, 233, 235
  professional recognition as artist, 226
  Rands, Angus, 224
  Ranson, Ron, 224–225
  teaching, 233–236, *236*
  'wet into wet' technique, 232–235, *233, 234*
Park District commission, 236–237, *237*
Parkinson Building, 91–94, *94*
Pasture Cottage

— 245 —

Burtons, 145–146
Dainton, Fred, 146
finding Woolwich Farm, 149
Kearby village, 141–147
moving from, 149
neighbors
    Crockatt, Helen, 143, 145
    'La Contessa,' 143
    Morgans, 144–145, *145*
    Youatt, Brian and Quita, 143
new Austin Mini, 147–148
purchase, 131
Sue's Gateways School friends, 145–147
winter time, 148–149
Peel Park, 6, 27–28
pets, early years
    Paddy, 17–19
    Polly parrot, 17–19
piano, 19–20
Pickles, Jack, 10
Planning Officer position, University Arts Block, 137–138
    Botany department, 134–135
    Chamberlin Plan, 151–180, *170, 171, 172*
    Friend, Louise, 101–102
    Great Hall remodel, 100
    Gregory fellows, 138–139
    House and Estates Committee, 101–102
    Library, 87–90, 170–172
    Man-Made Fibres Building, 135–137, 136, 137
    new Refectory, 94–98
    Parkinson Building, 91–94, *94*
    support staff, 101–104
politics of family members, 25–26
Polly parrot, 16–17
Poulson, John, 157–158
Powell, Geoffrey, 152, 172, 174
Priestley, J B, 25
Princess Royal, 100, 135
Prof. *See* Evans, Dr R H
professional recognition as artist, 226
pursuing PhD, 50, 57, 76, 80

R

radio, 19
Ramsden, Brigadier General, 66
Rands, Angus, 224
Ranson, Ron, 224–225
rationing challenges, 76–77, 91
Red Dove, 35, 38, 46, 47, 48, 52
Refectory, development of, 94–98
relatives. *See also* Grandfather; Smith, Edith; Wilson, Arthur J.
    Morley, Emma Wilson, 26–27
    Smith, Anne, 15–16, 24, 25
    Smith, Arnold, 15–19, 18
    Smith, Mabel, 15–17, 24, 25
    Wilkinson, Cecil, 22–26, 23
    Wilkinson, Gertrude, 22–25
    Wilkinson, Tom, 25–26
    Wilson, Eliza, 26–27
    Wilson, Pamela, 26–27
    Wilson, Polly, 27
retrievers
    Bridie, 210–211
    Minnie, 211–212, 228, 236
    Salty, 208–211
Richard III, King, 196, 197
Rider, Sidney, 73–74
Ridgeway, Hugh
    Marmolada, 113–119, *115, 116*
    Mont Buet, 105–113, 1*08, 109*
    project architect, 104
Roslyn Terrace, 3, 19

S

Salt, Sir Titus, 23–24
Saltaire, 23–24
scout troop. *See* Boy Scouts
sculling, 51–52, *52*
Second World War
    barrage balloons, 62, *63*
    Fighting Vehicles 9 department, 59, 61, 69
    London living arrangements, 59–60, 61, 66–69
    my invaluable bicycle, 60, 66, 70
    Ode of Remembrance, 69
    Ramsden, Brigadier General, 66

Index

Swingler, Mrs., 70–71
Tank Supply 9 department, 58–59
Toc H, 61, 66–69
Tube station air raid shelter, *60*
university studies, 57, 59
V1 flying bombs, 63–70, *64*
V2 rockets, 67, 67–68
volunteer duties, 57–58
war's end, 69–71
Whelan, Pat, 62, 63, 64–65, 69
Selby, Miss, 98–100
selling Woolwich farm, 227–228
Shaw, Malcolm, 6, 10, 21, 27–28
shopping
    Chicago relocation challenges, 229–231
    early years, 9–11
Singing Hawk, 35–36, *36, 37,* 38, 45
Skipton Castle, 197, *198*
Smith, Anne, 15–16, 24, 25
Smith, Edith. *See also* Wilson, Arthur J.
    chores, 7–9, *8*
    early years, 1
    High Wycombe holiday, 70–71
    master baker, 11, 13, 14, 46–47
    my son the thief, 39
    personality, 7
    politics, 25–26
    as secretary, 9
Smith, Mabel, 15–17, 24, 25
Somers, Lord, 44
Southern Hospital, Sweden, 163–164
Spence, Sir Basil, 139-140
stealing a Bunsen burner, 39
Stevens, Sir Roger, 12, 99, 160, *171*
Stockdale, John, 214
student housing
    Bodington hall development, 165–168
    'Henry Price' flats, 165
    Jones, Denis Mason, 165–170
    touring other countries' universities, 159–165
surveying
    in Scouts, 53, 54–55
    University years, Leeds, 77
Susan. *See also* Smith, Edith

    birth, 131, 133
    encouraging Chicago move, 226–231
    Gateways School friends, 145–147
    Morgans, 144–145, *145*
Sutherland, Dr., 78
Swingler, Mrs., 70–71
Switzerland scouting trip, 46–50, *49*

T

Tank Supply 9 department, 58–59
teaching, 233–236, *236*
tent feed, 38–39
Tetley, Brigadier, 102
Thirlwall, Geoffrey, 158–159
Toc H, 61, 66–69
    Ode of Remembrance, 69
topping out ceremony, 170
touring other countries' universities, 159–165
    Denmark, 162
    Finland, 159–161
    France, 164
    Leipzig, 192–195, *194*
    Norway, 163–164
    Sweden, 162–163
    Switzerland, 164–165
toys
    Flies Right Off the Ground aeroplane, 27
    model boats, 27–28
Trabant road trip, 193–194, *194*
tram travel, 29–30
TS9. *See* Tank Supply 9 department
Tsavo Game reserve, 190–191, *191*
Tube station air raid shelter, *60*
Twyford, Sir Henry, 44, 45

U

Ullmann, Herr Helmut, 194–198
Undercliffe, England, 5–6, 9, 14, 28
university Library troubles, 87–90
University of East Africa central catering unit
    arrival in Nairobi, 187–188
    Aston, Guyron, 188–191
    Glew, George, 186–187, 191–192

Jones, Royston, 188–191
  Principal report, 191
  related problems, 192
  Tsavo Game reserve, 190–191, *191*
University of Leipzig tour
  arrival in Berlin, 192–193
  Crawford, Ron, 192–198
  Hans the tour guide, 193–198
  reciprocal England visit, 195–198
  restoration sites, 194–195
  Trabant road trip, 193–194, *194*
  Ullmann, Herr Helmut, 194–198
University years, post-war 1945–1978
  architect credentials, 78–81
  Brown, Bursar, 100–101
  Brown, James, 84–85
  Brownlie, Miss, 97
  Butler, Reg, 138–139
  Chamberlin Plan
    accolades, 173–174
    BDP consultants, 174–175, 179, 180
    Bon, Christopher, 152, 172, 174
    Central Electricity Generating Station, 179–180
    Chamberlin Development Plan Book, 153–156
    Chamberlin's partners, 152
    Chamberlin's personality, 152–153, 173
    cost control measures, 173
    flexibility in building uses, 154
    Kaberry, Sir Donald, 174–176
    lecture theatres, 155–156, *171*
    library, 170–172
    Mathematics and Sciences building, 170, *172*
    medical school, 174–178
    Poulson, John, 157–158
    Powell, Geoffrey, 152, 172, 174
    ring road proposal, 154, 156, 158–159
    selection for T A Lodge replacement, 151–152
    student housing, 159–168
    Thirlwall, Geoffrey, 158–159
    topping out ceremony, 170
  touring other universities, 159–165, 177–178
  Wood, Derek, 177–178
  Civil Engineering department development
    Architectural Engineering course, 82, 139–140
    Chapman, 77–78
    designing new Engineering building, 78–80
    Evans, Dr R H, 73–75, 78–79, 81
    Happold, Ted, 82
    Holford, Sir William, 140–141
    lecturer duties, 74–75, 77, 80–82
    Lodge, T A, 79–80, 91, 93–94
    Maillart, Robert, 81
    Morris, Vice Chancellor Charles, 79, 79n2, 80, 82, 86, 91, 100, 102–104
    rationing challenges, 76–77, 91
    Rider, Sidney, 73–74
    Sutherland, Dr., 78
  Denny, James, 92–93
  Devonshire Hall residence, 83–86
  discordant dueling pianos, 84–85
  Duchess of Kent, 100, *101*
  Hilliard, Hall Porter, 98–99
  Ilett, J J, 86–87, 101
  Lister, Miss, 133–134
  Manton, Irene, 134–135
  married life
    engagement and wedding, 126–127
    Majorca holiday, 128–131
    pregnancy, 127–131, 133
    search for a home, 128, 131
  missed dinner engagement, 83–84
  Ogilvie, Lady, 99
  Pakistani immigrant ingenuity, 90
  project architects
    Chamberlin, Peter, 152–174, *170, 171, 172*
    Johnson, Allan, 80, 104, 137, 179
    Jones, Denis Mason, 100, 103
    Knighton, A L, 102–103
    Lasdun, Denys, 151, 174
    Lodge, T A, 79–80, 91, 93–94, 103–104, 135–137

Ridgeway, Hugh, 104
pursuing PhD, 76, 80
Royal Institute of British Architects, 79n1, 83, 162, 172n2
Selby, Miss, 98–100
student housing considerations
    Bodington hall development, 165–168
    'Henry Price' flats, 165
    Jones, Denis Mason, 165–170
    touring other countries' universities, 159–165
Tetley, Brigadier, 102
University Planning Officer position
    Arts Block, 137–138
    Botany department, 134–135
    Chamberlin Plan, 151–180, *170, 171, 172*
    Friend, Louise, 101–102
    Great Hall remodel, 100
    Gregory fellows, 138–139
    House and Estates Committee, 101–102
    Library troubles, 87–90
    Man-Made Fibres Building, 135–137, *136, 137*
    new Refectory, 94–98
    Parkinson Building, 91–94, *94*
    support staff, 101–104
Williams, Sir Owen, 80–81, *81*
Williamson, Edmund, 167
winter driving, 82–83, *83*
Woodsley Hall
    Brittain, Dr. Robert, 123, 124–125
    Ellis, Stanley, 124
    Fawcett, Eleanor Margery, 121–123, *122*
    hall raids, 126
    Wardenship, 119–123, *122*
    Woodhead, John, 123
University years, pre-war, 50, 52–53
    civil engineering, 50, 52–53
    Evans, Dr R H, 50
    Garton, Sam, 51, 53
    qualifying for university, 50
    school scout troop
        bridge building, 52–53

    fog patrols, 53–54
    sculling, 51–52, *52*
    social life, 51–52
    surveying, 53, 54–55

V

V1 flying bombs, 63–70, *64*
V2 rockets, *67,* 67–68
Valley View Grove, 21, 30
VE day. *See* war's end

W

war's end, 69–71
'wet into wet' technique, 232–235, *233, 234*
Whelan, Pat, 62, 63, 64–65, 69
Wilkinson, Cecil, 22–26, *23*
Wilkinson, Gertrude, 22–25
Wilkinson, Tom, 25–26
Williams, Sir Owen, 80–81, *81*
Williamson, Edmund, 167
Wilson, Arthur J. *See also* Smith, Edith
    Askrigg holiday, 196–197
    boiler treatment business, 4–5
    Brownie camera, 47
    first jobs, 2–4, *3*
    first radio, 19
    High Wycombe holiday, 70–71
    ice cream, 15
    marriage advice, 126
    motto, 4
    'Mystery Tours,' 2–3, *3*
    personality, 7
    tram ride, 29–30
    World War I, 1
Wilson, Eliza, 26–27
Wilson, Pamela, 26–27
Wilson, Polly, 27
winter sports, 28
Wood, Derek, 177–178
Woodcraft Circle, 35, 37–38
Woodhead, John, 123, 127
Woodsley Hall
    Brittain, Dr. Robert, 123, 124–125
    Ellis, Stanley, 124
    Fawcett, Eleanor Margery, 121–123, *122*

hall raids, 126
Wardenship, 119–123, *122*
Woodhead, John, 123
Woolwich farm
  Arthur, 206–207
  barn build, 216–218, 218, *220*
  Barrett family, 201, 205–207, 215, 216
  caravan park proposal, 200–201
  first seeing, 149
  Foxton, Barry, 202–204
  Granville, 206–207, *209*
  haymaking, 214–216
  horseback riding, 212–214, *213*
  initial condition, 199–200, *200*
  Mancia, 212–214, *213*
  Mandy, 212–214
  Pearson, Mr., 215, 216
  pheasant hunting, 208–210, *209*
  remote country living challenges, 219–222, *221*
  renovation, 201–204, *205*
  retrievers, 208–212
  Scout visits, 218–219
  selling Woolwich farm, 227–228
  winter scene, *207*
  village friends, 205–208

Y

Yeadon aerodrome, 13, 21, 27, 55
Youatt, Brian and Quita, 143